Refining Used Lubricating Oils

Refining Used Lubricating Oils

Editor

Preeti Dhama

scitus
academics

Refining Used Lubricating Oils

Edited by **Preeti Dhama**

Printed in 2017

ISBN: 978-1-68117-034-3

Library of Congress Control Number: 2015931836

Contents

vi

Preface

Re-refining used oil restores the chemical composition of the base oil so that it can be used to produce new lubricant products over and over again. Re-refining is an energy efficient and environmentally beneficial method for managing used oil. Instead of burning the used oil, which releases harmful emission into the atmosphere, re-refining conserves the base oil.

Technological advancements for re-refining used oil since the mid-1990's, have resulted in superior quality base oils and lubricants in the market today that can meet the most stringent standards for automotive engine oil specifications. Because of the indefinite lifecycle of re-refining used oil, substantial benefits are achieved with respect to environmental impact, reduced energy required for manufacturing lubricants, and conservation of a non-renewable petroleum resource.

Refining Used Lubricating Oils focuses on the properties of used lubricating oils and presents methods these materials can be re-refined and converted into useful lubricants along with other products. It gives an up-to-date review of most of the processes for used lubricating oil refining that have been proposed or implemented in different parts of the world, and addresses feasibility and criteria for selecting a particular process.

Editor

Chapter 1

Fundamentals of Lubricants and Lubrication

Walter Holweger[1]

[1] Schaeffler Technologies AG & Co.KG, R&D Central Materials, Germany

INTRODUCTION

Literature about lubricants is available in all public domains. Readers should search at those platforms in the case of special interests. Citations given here do not represent the full scale but reflect an overview from a today's perspective. [1-7]

Part of this chapter will be the basic chemical structure of lubricants including some property descriptions. Since literature in tribology is innumerous, the reader should check his special area of interest.

Lubricants play a key role in machinery element safety. Their main tasks are:

- to keep moving parts apart from each other,
- to take heat out of the contact by their through pass,
- to keep surfaces clean,
- to transport functional additives toward the surface and
- to transfer power in the application (hydraulic, automatic transmission, breaks). [6, 8]

Functionality of lubricants is defined by their chemical structure and their physical properties. Basics of lubrication are covered by organic chemistry to a major and inorganic chemistry to a minor extent. [2,3]

Lubricants are regulated internationally and locally, e. g. by ASTM (American Standard of Testing Materials) or DIN (Deutsche Industrienorm). Regulation covers the physical, chemical and toxicological description of lubricants including safety guide lines and others. [2, 3]

SOME BASICS

The spatial structure of carbon chemistry defines all activities of the lubricants derived from them. The spatial structure of organic carbon chemicals is given by the binding state of carbon. [10]

Three main types are discussed. Two are essential for lubrication: single and double bonds.

Single Bonds: Tetrahedral Binding

In the tetrahedral binding state, reflecting the status of single bonds, carbon is placed in the centre of a pyramid with bindings into space from the centre to the corner (Figure 1).

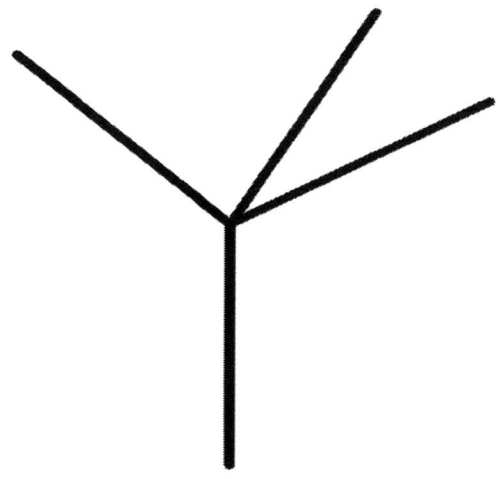

Figure 1: Tetrahedral binding of carbon.

Carbon is placed in the center of the tetrahedral with four attached valences. Within chemical convention in order to abbreviate the structure denotation the atom symbols are neglected.

Carbon may bind to another one by corner to corner. (Figure 2)

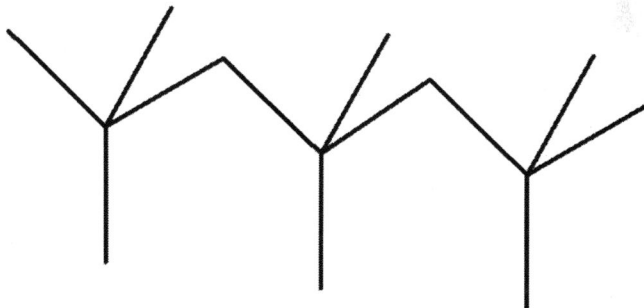

Figure 2: Corner to corner binding state.

Corner to corner binding leads to zigzag chains, where the angle of carbon to the hydrogen atoms is 108°. In general the hydrogen is neglected, leading to a skeleton drawing of the structure.

Beyond the fixed angle of 108° and the zigzag shape of such hydrocarbon structures, a high variety of structures arise due to the fact that those bindings may branch or bind to cyclic structures. (Figure 3 and Figure 4)

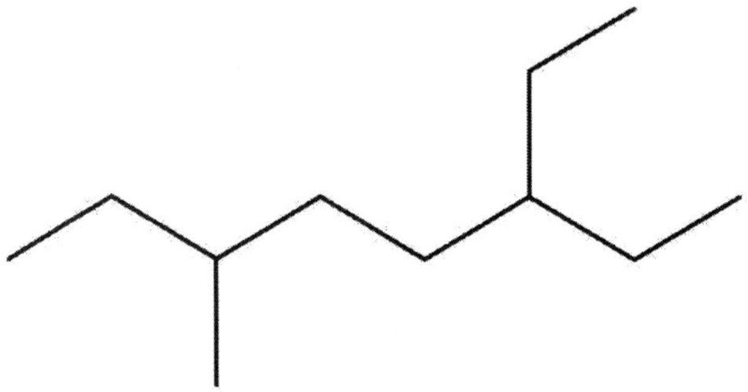

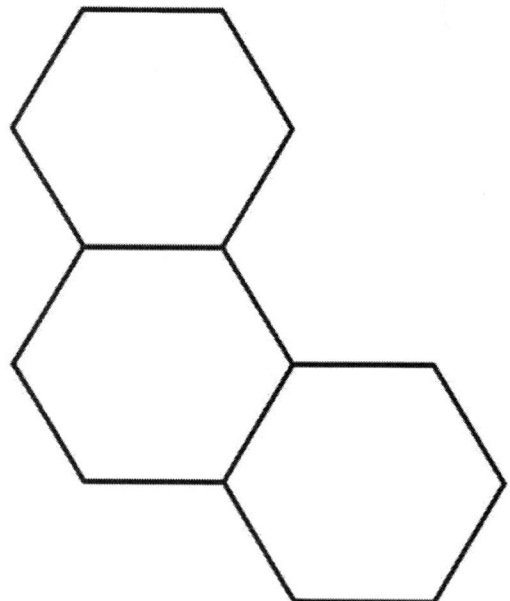

Figure 3: Branched structures by carbon to carbon binding.

Figure 4: Cyclic Structures by carbon-carbon binding.

Single bonds in hydrocarbons are free to rotate (Figure 5). Rotation leads to the situation that hydrogen atoms within the chain get close to each other. As a consequence the energy of the molecule rises.

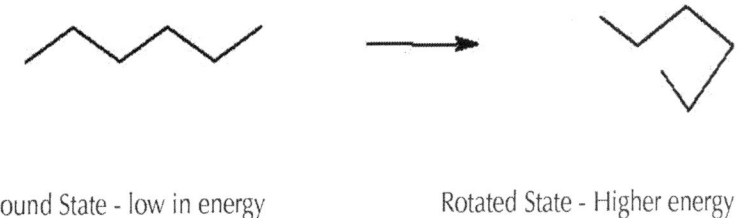

Ground State - low in energy   Rotated State - Higher energy

Figure 5: Energy rise in rotated structures.

Similar to internal rotation, molecular energy rises if molecules get under stress by moving them closely together without giving time to relax. (Figure 6)

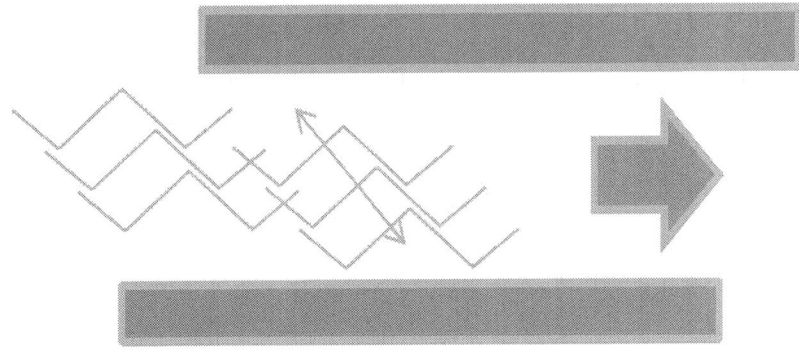

Figure 6: Excitation by pushing molecules to one another by shear stress.

Also the fact of putting or pressing molecules toward a surface may lead to a steep increase in internal molecular energy, sometimes high enough to cut them.

Double Bonds

Carbon may also bind to others by double bonds, such that two of the four bindings attach to the other as double, whereas the remaining bonding stays single. (Figure 7)

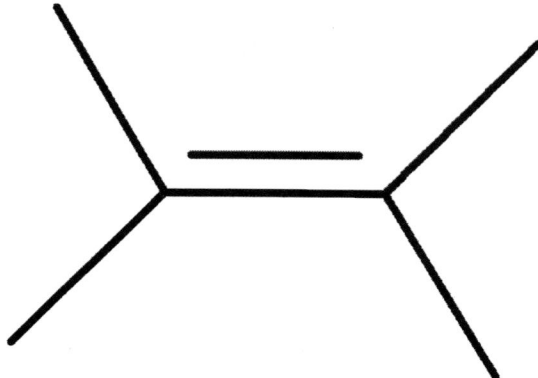

Figure 7: Double bonding.

Double bond shows a 120° neighborhood angle to the carbon. This angle is kept constant and will lead toward different structures in the double bond chemistry. (Figure 8)

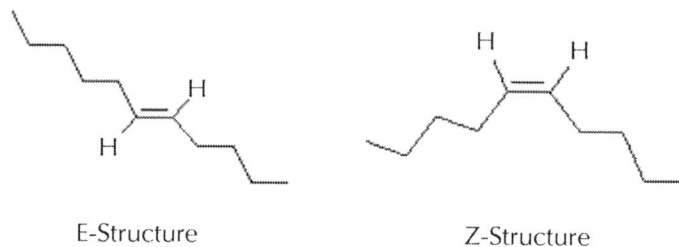

E-Structure   Z-Structure

Figure 8: E and Z structures in double bond.

Both structures differ in their energy. Double bonds are part of biodegradable additives (native oils) but also additives and thickeners in the case of greases. Z-Structures are dominant in native oils.

Triple Bonds

Triple bonds are seldom found in tribology. They represent a high energy state in molecules with very high reactivity. As such, they are part of catalytic degradation processes in lubricants. Within a triple bond carbons attach to each other by a linear principle (Figure 9):

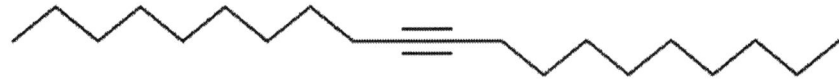

Figure 9: Triple bond present in a hydrocarbon.

BASE OILS IN LUBRICATION: GENERAL COMMENTS ABOUT SPECIE AND GROUPS

Hydrocarbon Base Oils for Lubrication derive from organic chemistry. Different categories are given by their chemical composition and structure. [2]

Hydrocarbons, e.g. Structures that contain solely Hydrogen and Carbon (H, C)

Ester Oils, e.g. Structures containing Hydrogen, Carbon and Oxygen. Some Esters are derived from other precursors, such as phosphoric acid esters.

Polyglycoles: Structure containing Hydrogen, Carbon and Oxygen but being different in binding state compared to Esters.

Within a general scheme, base oils are identified as Groups.

- *Group I*: Those lubricants are built from saturated hydrocarbons, e.g. hydrocarbons without alkenes (hydrocarbons with double bonds)) (> 90%), obtained by solvent extraction processes and catalytic hydrogenation. Sulfur may part in amount of > 0.03%. Viscosity index (VI) is in between 80 and 120.
- *Group I+*: Oils that are in a VI range of 103-108.
- *Group II*: Hydrogenated (saturated) hydrocarbons (> 90%) and sulfur below 0.03% per weight with viscosity index (VI) of 80 till 120.
- *Group II+*: Oils in the VI range of 113-119.
- The base oil within this group is manufactured by hydrocracking, solvent extraction or catalytic dewaxing processes. Those oils are pale or water like colored.

- *Group III*: Oils with a saturation > 90%, sulfur < 0.03% and a viscosity index > 120. Those oils are produced by catalytic procedures with a concurrent rearrangement of the carbon backbone during hydrogenation.
- *Group III+*: Oils providing a VI at least of 140.
- *Group IV*: Poly-α-Olefins with sulfur content approximately 0, viscosity index 140–170, being produced by catalytic polymerization of low molecular weight end terminated olefins.
- *Group V*: All other oils, e.g. esters, polyglycoles, phosphate esters.

North America states Group III, IV and V as synthesized hydrocarbons (SHC) while in Europe Group IV and V is declared as synthetic oil.

SATURATED NATURAL HYDROCARBONS

Saturated hydrocarbons are those who do not contain double bonds in their structure. They derive from the tetrahedral binding of carbon (bindings that point into corners of tetrahedron). The simplest structure is given by methane, ethane, propane, butane with carbons attached at the corners of the tetrahedral. These representatives are present in the natural gases, while methane is found in enormous quantities as methane-ice cluster. The gases themselves are not in use as lubricants but are components of fuels. (Figure 10)

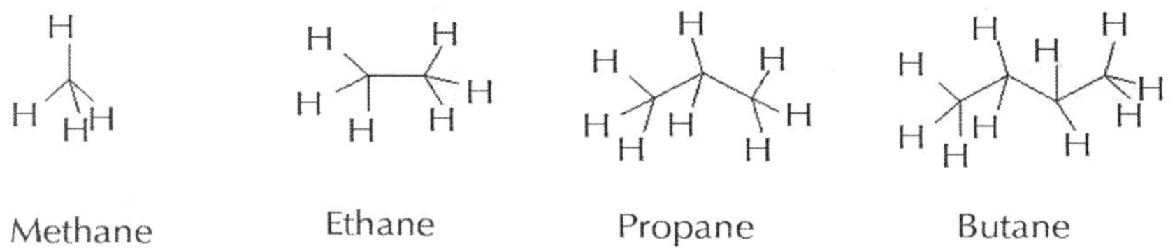

Methane   Ethane   Propane   Butane

Figure 10: Methane, Ethane, Propane, Butane.

Starting from pentane the hydrocarbons get liquid and are the principal components of fuels, solvents, and raw materials for the chemical industry. To facilitate reading and drawing only the carbon backbone is drawn without explicitly showing hydrogen. (Figure 11)

Figure 11: Pentane, Hexane, Heptane.

Binding of carbon to carbon may be realized in chains, but also in branched chains and different cycles (Figure 12).

Figure 12: Methylbutane (Isopentane), 2,2-Dimethylpropane (Neopentane), Cyclohexane.

Hydrocarbons from C10 on till C14 are in use as solvents for cleaning (C11-C13 isoparaffines) (Figure 13):

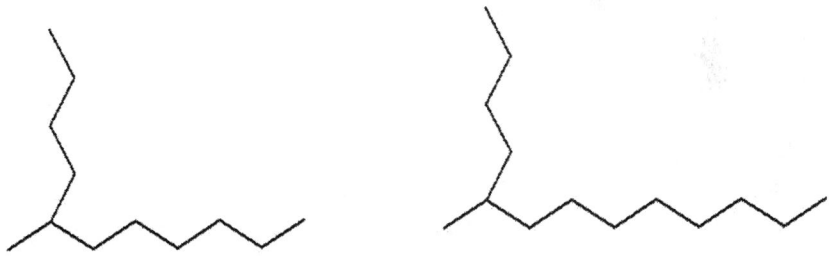

Figure 13: C11-C13 Iso paraffines.

From C16 on, hydrocarbons represent typical structures present in lubricants. As the linear hydrocarbons, beginning at C18 are solids, they are common in waxes and thickeners for liquid hydrocarbons. Due to their high solidification point they are a threat if present in Diesel fuels by blocking filters.

Apart from their function as hydrocarbon waxes they are not suitable as lubricants for machine oil circuits.

Suitable lubricants are derived from C16–C70 hydrocarbons with branched chains. Branching leads to low pour points (the point where the lubricant starts to get solid). Machine oils with low pour point, suitable for low temperature applications are branched in their carbon chain. (Figure 14)

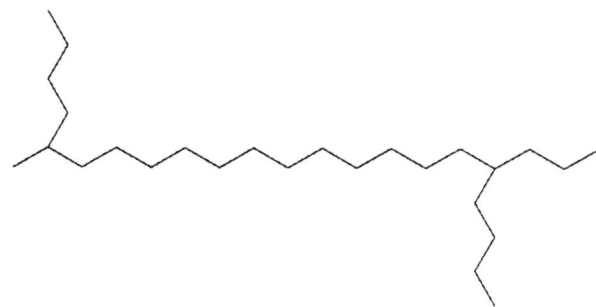

Figure 14: Representatives of saturated hydrocarbons as typical lubricants.

In general, the viscosity of a lubricant - as a measure for the ability to move across - increases with the molecular weight, e.g. the number of carbon atoms attached. Viscosity is measured by different techniques. Basically the lubricant is pushed or moved in between plates or by moving it in the gravity field. International convention states 16 classes of viscosity as an ISO Standard (ISO VG classes): ISO VG 5, 7, 10, 15, 22, 32, 46, 68, 100, 150, 220, 320, 460, 680, 1000 and 1500. Low numbers indicate low viscosity, higher numbers high viscosity. Since viscosity is strictly related to temperature, the ISO VG classification refers to 40°C as a standard temperature. The nature of measuring the viscosity leads to the physical value of an area per time: mm^2/s. Hence, ISO VG 68 for example denotes a viscosity of the lubricant, measured at 40°C within 68 mm^2/s within a range of roughly 10% below and 10% beyond the given 68mm^2/s.

Low molecular weight, branched hydrocarbons are often used in *pneumatic spraying*, due to their viscosity range, starting at 2 (water-like), 5, 10 and 15.

Low viscous hydrocarbons from ISO VG 10, 15, 22, 32, 46, 68 and 100 are in use as *hydraulic oils.*Common hydraulic oil viscosity is around ISO VG 32, 46 and 68.

Hydrocarbons with higher viscosities are part of *machine oils*, carrying out the ordinary lubrication functions. Machine oil viscosities are in the range of ISO VG 68, 100, 150, 220, 320, 460. The number of carbons is in the range of 30–80 in the chain.

Some applications in heavy duty processes demand viscosities even higher in the range of ISO VG 680, 1000 and 1500.

Cyclic Hydrocarbons (Naphtenes)

Naphtenic hydrocarbons are derived from hydrocarbon cycles with more or less long chains attached to the cycle. Due to their high branching they are very common in low temperature applications (below -30°C) for hydraulics; low temperature greases. (Figure 15)

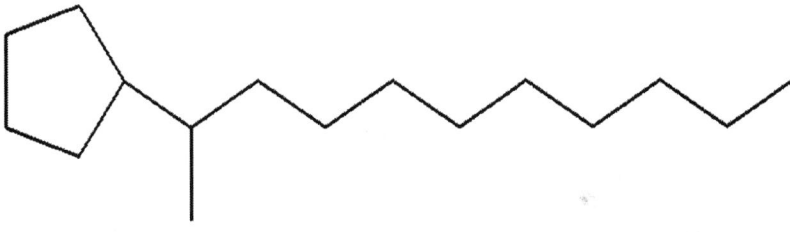

Figure-15: Principal Structure of Napthenic Hydrocarbons.

Aromatic Hydrocarbons (Alkyl Aromats)

Aromatic Hydrocarbons (Alkyl Aromats) derive from the six-membered benzene ring system, attached by hydrocarbon chains. Aromatic hydrocarbons are in use for low temperature applications.

Alkyl Naphthalenes are a modern group of aromatic hydrocarbons. They may act as solvent improvers for synthetic oils, facilitators in generating greases, low temperature applications and much more (Figure 16):

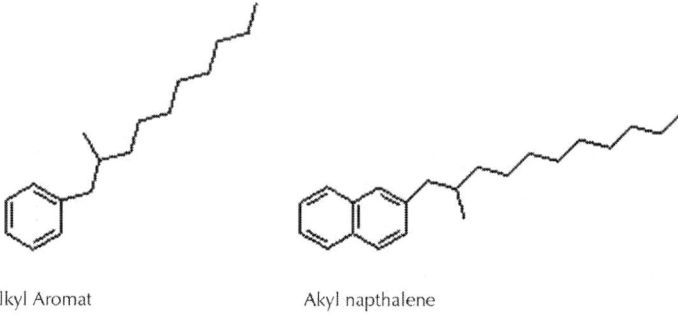

Alkyl Aromat   Akyl napthalene

Figure 16: Alkyl Aromats and Naphtalenes.

Aromats and aromat-containing hydrocarbons are very vulnerable toward oxygen. Oxidation of aromats starts at the attached hydrocarbon chain, proximate to the aromat nucleus by a radical attack. This position is always very sensitive in similar structure, due to the fact, that the intermediate carbon radical is stabilized by the aromat and thus starts to stay persistent. As a fact, aromats may strongly boost oxidation of hydrocarbons if present in the mixture due to the mentioned persistency of the reactive intermediates. (Figure 17)

Aromats and naphtenics (containing unsaturated hydrocarbons and aromats) should be stabilized against oxidation.

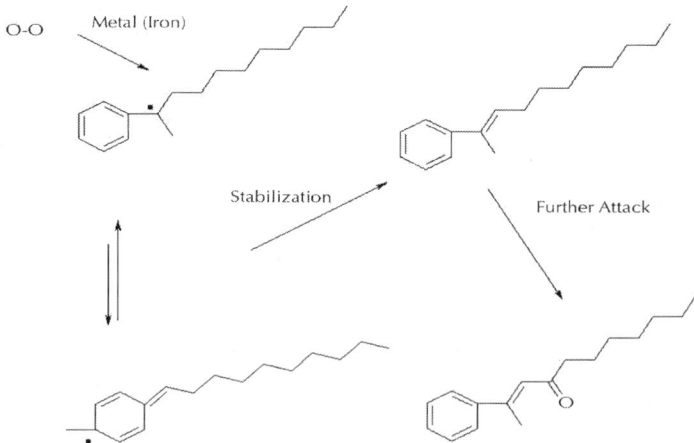

Figure 17: Oxygen attack in the oxidation mechanism of Alkylaromats.

SYNTHETIC HYDROCARBONS

Poly- -Olefins (Pao)

PAO is dominating all synthetic hydrocarbons by amount of production and worldwide turnover. Syntheses start from Dec-1-ene, a linear C10 hydrocarbon with a double bond at the beginning of the molecule. Polymerization and hydrogenation leads to PAO, as a highly branched and fully saturated hydrocarbon (Figure 18). [2, 4]

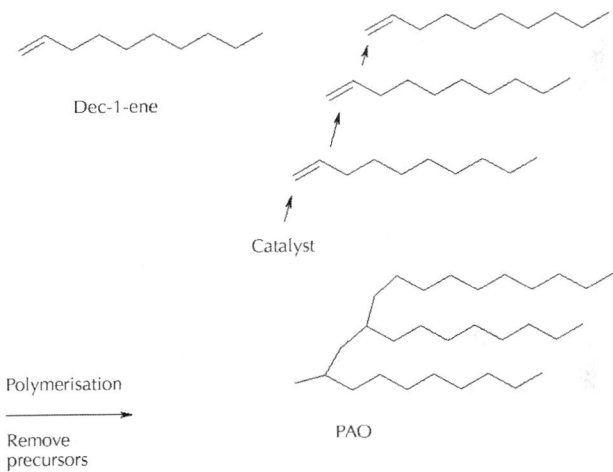

Dec-1-ene

Catalyst

Polymerisation

Remove
precursors

PAO

Figure 18: Principle of PAO formation.

Modern PAO may also start from a variety of hydrocarbons (C8-C12) by the same processes. PAO are the most prominent worldwide used hydrocarbons and found in all important applications, e.g. gear oils, circuit oil, hydraulic oil, base stock for automotive applications and others. [2, 3, 6, 7]

The extraordinary importance of PAO is due to its applicability at very low temperatures (pour points below –30°C) and, in the case of suitable antioxidant prevention also at higher temperatures (> 120°C). While PAO is, by its structure, very common in low temperature applications, it is very poor in the contact with metal surfaces beyond 120°C if not properly additivated by antioxidants.

Principal antioxidants for PAO are Phenyl- -Naphtylamine (PAN) and octyldiphenylamines (see antioxidants (AO)).

Polyisobutenes (PIB)

PIB are a sub class of polymerized olefins. They are widely used to boost low viscous oils to higher ISO VG grades or as functional additives to improve the viscosity index (VI): the attitude of the lubricant to lower its viscosity strongly by temperature is reduced by addition of PIB. Synthesis is carried out starting from isobutene by catalytic oxidation processes (Figure 19):

Isobutene   Polyisobutene

Figure 19: PIB formation by catalytic polymerization of Isobutene.

Sulfurization with activated sulfur precursors lead toward sulfurized isobutenes (SIB) widely used as extreme pressure (see also section about EP/AW additives).

ESTER OILS

General

Esters are in general reaction products between alcohols and acids. Their formation is also possible by means of other techniques, e.g. specific oxidation reactions, rearrangements in organic molecules or different reactions. [10]

Carboxylic Acid esters are created by the reaction of alcohols and carboxylic acids [A] and their derivatives, by trans–esterification (B), or catalytic reactions, e.g. epoxides with carbon dioxide (C) (Figure 20). [10]

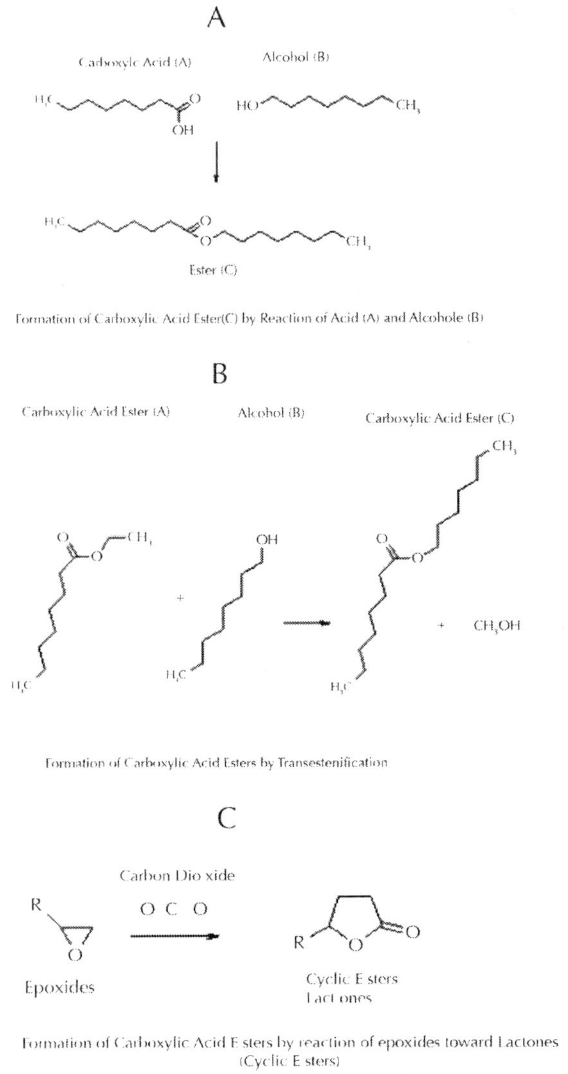

A

Carboxylic Acid (A)   Alcohol (B)

Ester (C)

Formation of Carboxylic Acid Ester(C) by Reaction of Acid (A) and Alcohole (B)

B

Carboxylic Acid Ester (A)   Alcohol (B)   Carboxylic Acid Ester (C)

+   →   +   CH$_3$OH

Formation of Carboxylic Acid Esters by Transestenification

C

Carbon Dio xide

O C O

Epoxides

Cyclic E sters
Lact ones

Formation of Carboxylic Acid E sters by reaction of epoxides toward Lactones (Cyclic E sters)

Figure 20: Examples for creation of esters. (A) Reaction of Carboxylic Acids with Alcohols, (B) Transesterfication, (C) Reaction of Expoxide to cyclic Esters.

Esters in Lubrication Technology

Despite their high variety in structure esters are used in different categories: [1, 2, 4]

Mono-Esters

Mono-Esters derive from a monocarboxylic acid (Carboxylic Acid that contains only one acidic centre) and monofunctional alcohols (Alcohole with only one OH group). [10]

Esters derived from this structure are seldom used as pure lubricants, more as solvents or dispersants. For example alcohol ethoxylates, formed by addition of alcohols to epoxides may be esterified by a monocarboxylic acid leading toward a dispersant or self-emulsifying solvent. (Figure 21)

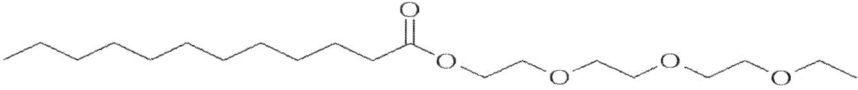

Figure 21: Mono Ester Formation with the specialty of esterified alcohole ethoxilates.

Di-Esters

Di-Esters are synthesized by use of dicarboxylic acids, mainly adipaic or sebacaic acid and two molecules of an alcohole. 2-Ethylhexylalcohole (Iso Octanole), leading to Di-isooctyladipate (DOA) or Di-isooctylsebacate (DOS, DEHS = Di-ethylhexylsebacate). (Figure 22)

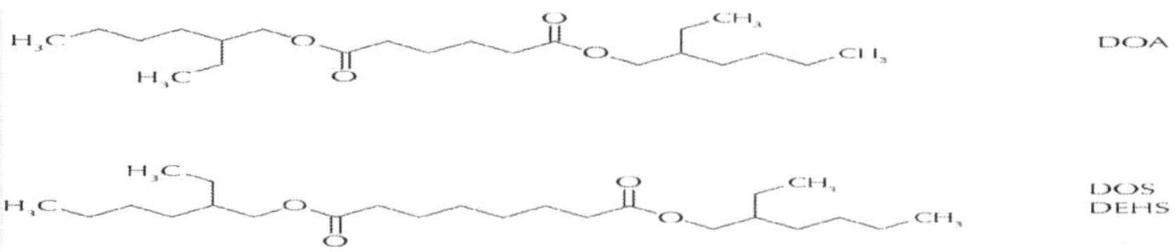

Figure 22: DOA and DOS.

They constitute an important group of oils, with either the function of base oil by themselves but also as adjuvant to mineral oil or PAO formulations.

For Di-Esters the reaction of alcohols (A) with two hydroxyl groups and a monofunctional carboxylic acid (B, B′) is also applicable. (Figure 23)

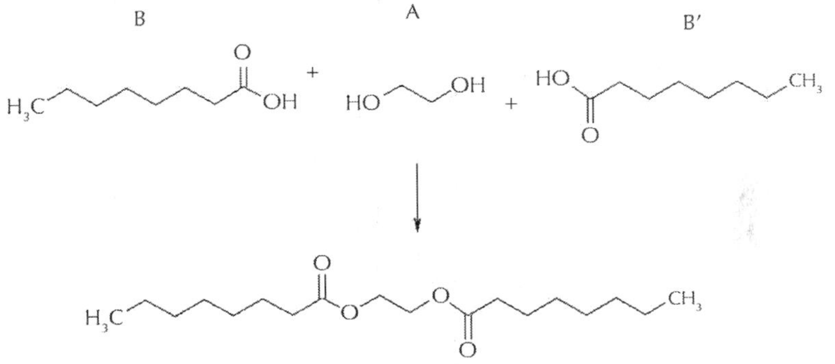

Figure 23: Formation of Di-Esters by Di-Alcoholes (A) reacted with Monocarboxylic Acids (B, B′).

For technical purposes the reaction product of Neo Pentylgylcole (3.3 Dimethyl-propane-1.4-diol) with oleic acid is important in lubrication technologies for use as a friction reducer and in minimal lubrication systems. (Figure 24)

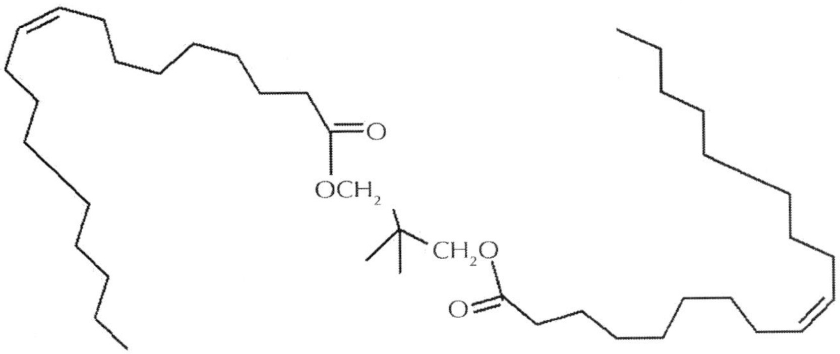

Figure 24: Neopentylgylcoledioleate (NPG-Dioleate).

Tri-Esters

Tri-Esters are mainly created by the reaction of trivalent alcohols with monocarboxylic acids. They are mainly represented by to major groups:

Glycerol Esters

Esters derived from glycerol as a trivalent alcohole leads to tri-Esters. (Figure 25)

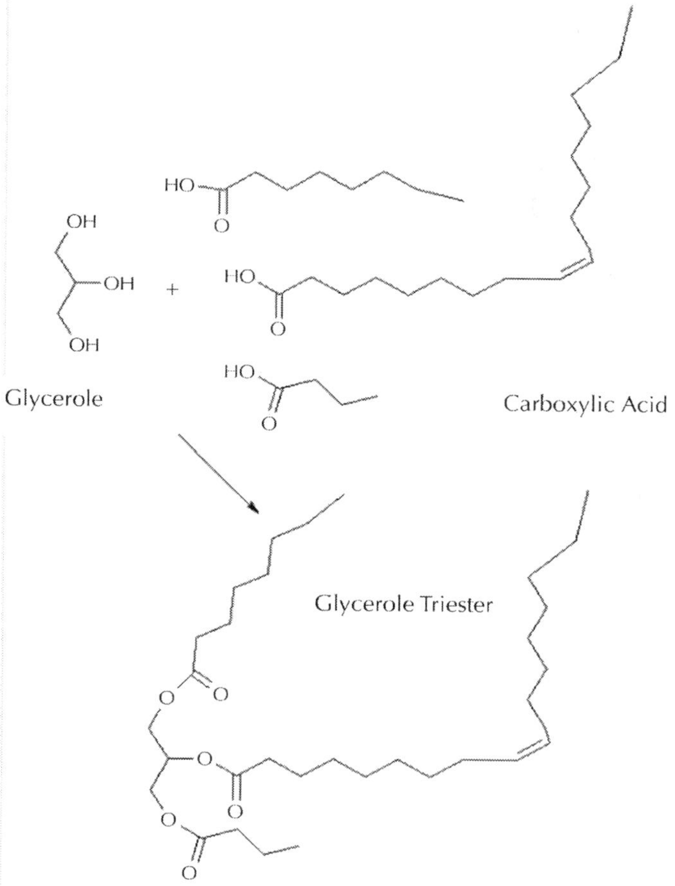

Figure 25: Esterification of Glycerole to tri-Esters.

Glycerol Tri-Esters represent the huge group of natural oils. Sunflower, rapseed oil are prominent representatives. A mixture of short chain carboxylic acids with unsaturated long chain acids is used.

As a fact of the presence of short chain carboxylic acids those esters are nutrients, biological degradable and widely used as natural, biodegradable oils.

As a special glycerol ester, important for lubrication, ricinoleic acid esters have to be mentioned.

Within this group ricinoleic acid represents the group of 12-hydroxy substituted C18 carboxylic acids.

Hence, alkaline cleavage of ricinoleic acid glycerol esters lead to 12-Hydroxi-oleic acid on the one hand and to sebacaic acid on the other hand by degradation of the double bond.

Catalytic hydrogenation of 12-Hydroxioleic acid results in the formation of 12-Hydroxistearic acid, which is important for modern grease concepts. Sebacaic Acid on the other hand is a raw material for DOS (see above) but also for the production of complex greases. (Figure 26) [1, 2, 4]

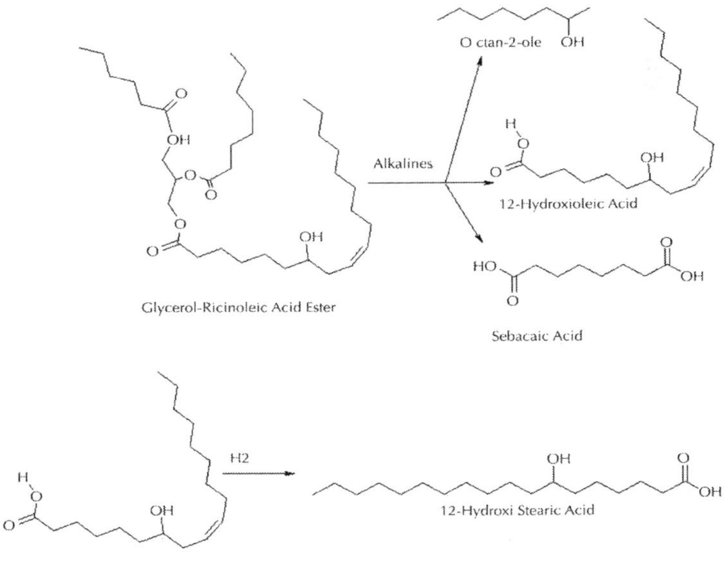

Octan-2-ole

12-Hydroxioleic Acid

Sebacaic Acid

Glycerol-Ricinoleic Acid Ester

Alkalines

H2

12-Hydroxi Stearic Acid

Figure 26: Cleavage of Glycerol – Ricinioleic Acid and hydrogenation to 12-Hydroxistearic Acid, Sebacaic Acid and Octan-2-ole [10].

Glycerole Esters with long chain carboxylic acids only, e.g. Glycerole Tristearate, are no longer nutrients and sparingly biodegradable. They are used as emulsifiers, consistency givers.

Glycerole Trioleate is a powerful friction reducer in tribological applications.

Triesters, Derived From Alcohols Else Than Glycerole

Trimethylolpropane Esters (Tmp-Esters)

TMP-Esters are created out of Trimethylolpropane (TMP) by reaction with short chain carboxylic acids, e.g. the range from C6 to C10. (Figure 27)

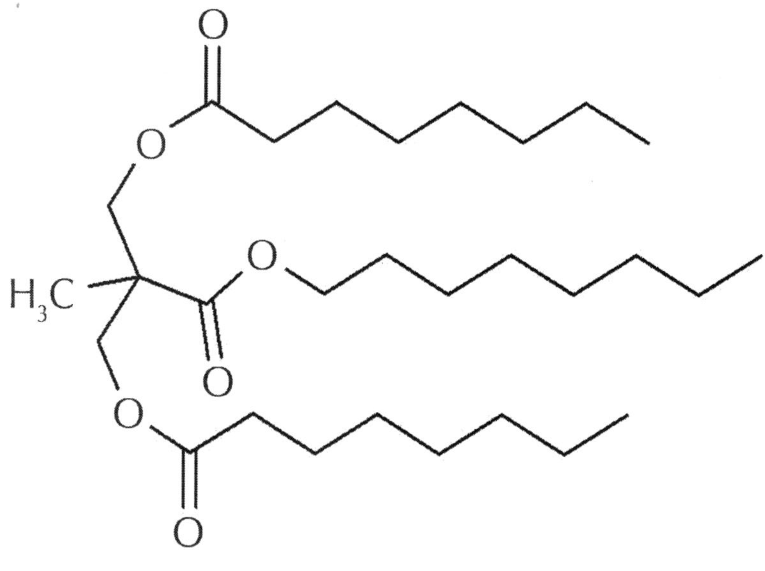

Figure 27: TMP Esters.

TMP Trioleate is created by reaction of TMP with oleic acid or by trans-esterification.and commonly used as lubricant in minimal lubrication.

Trimellitic Esters (Tm-Esters)

Apart from the described structures where trivalent alcohols get reacted with monocarboxylic acids, trimellitic Esters (TM-Esters) are products from Trimellitic Acid Anhydride with Mono alcohols. (Figure 28)

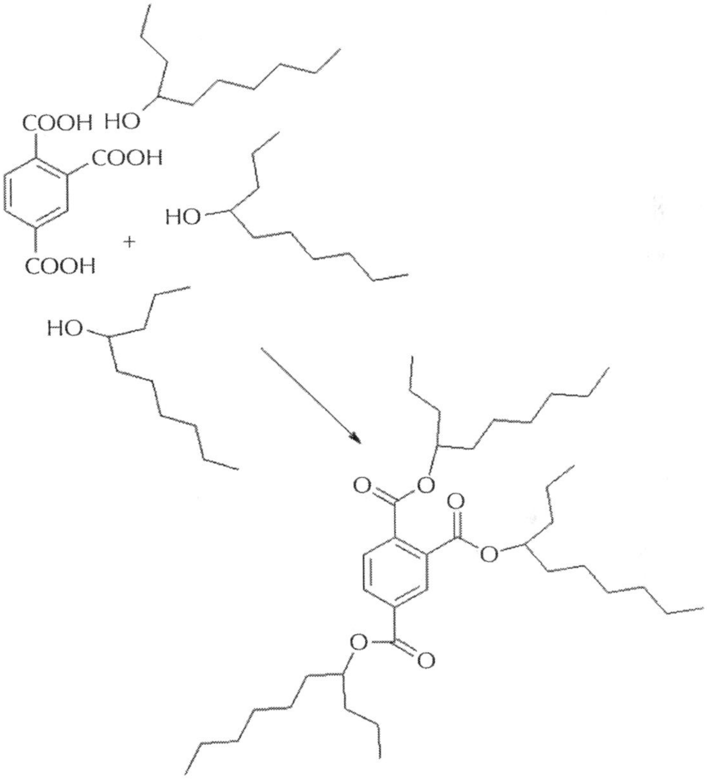

Figure 28: TM Esters by reaction of trimellitic acid with branched alcohole.

Due to the aromatic core those esters are high in thermal stability and widely used in high temperature applications.

Tetra Esters (Pentaerythrolesters, Pe-Esters)

Pentaerythrole acts as a four-valent alcohole which may be esterified by four carboxylic acids. (Figure 29)

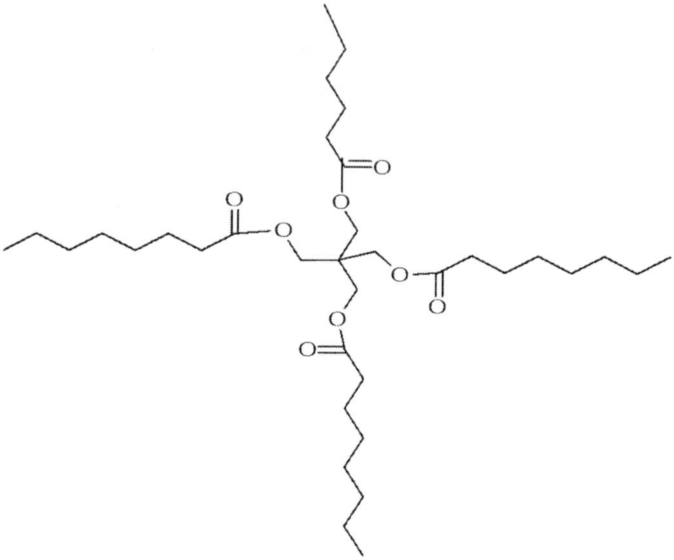

Figure 29: PE Esters.

Carboxylic acids are in the range from C6 to C10.

Dipentaerythrol Esters (Di PE Esters) are formed starting from Dipentaerythrole as a six-valent alcohole reacted by six monocarboxylic acids in a Carbon Chain length from 6 to 10. (Figure 30)

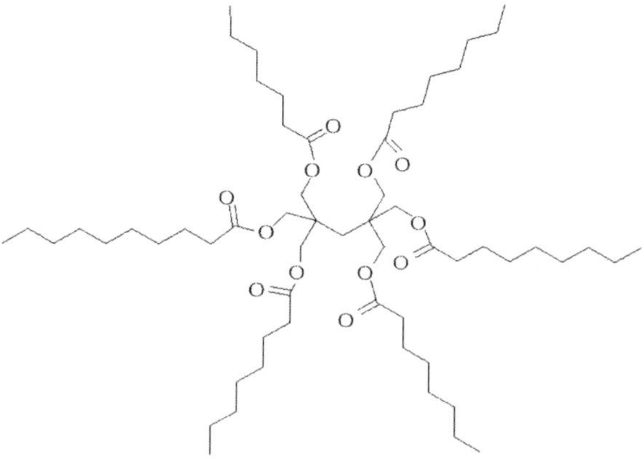

Figure 30: Di PE Esters.

Polyesters

In the past 20 years new groups of esters have been created by reaction of polycarboxylic polymers with alcohols. Those are reaction products of maleic acid anhydride (MSA) by Ene-Reaction with PAO precursors, leading to the PAO-backbone MSA addition product that might be esterified by butanole, leading to carboxylic complex esters (Figure 31).

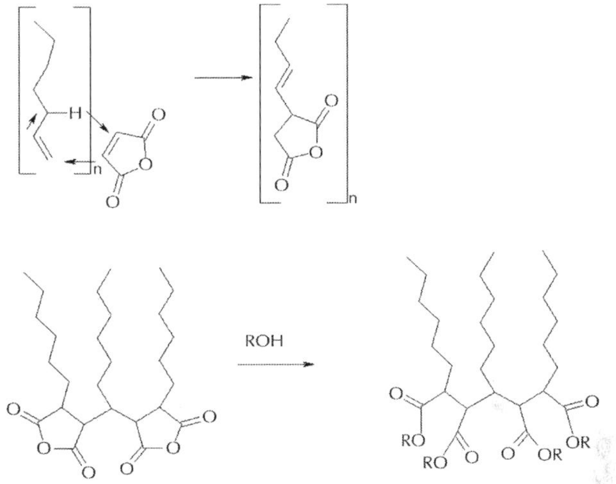

Figure 31: Complex Ester Formation by Ene-Reaction Sequences.

Complex Esters from those structures are widely used to improve the additive solubility and performance. Their structure with shielding the carboxylic groups causes less aggressiveness toward sealings.

STRUCTURE ACTIVITY RELATIONSHIP IN ESTERS

Esters are prominent representatives of lubricants where the chemical structure promptly leads to a specific tribological activity. However, if a tribological acitivity is demanded, the specific construction of esters may offer the solution.

Polar Acitivity

Esters are polar by their nature due to the central element where a carboxylic acid tail binds toward an alcohol. Polarity gives some advantage but also disadvantage in the case esters are used. In general, esters enhance the solubility of functional additives and keep them away from fall-out. Esters also enhance the cleaning of metal surfaces in operation, preventing a formulation by creation of sludge. Esters are, as a fact of their polarity, aggressive toward sealings with a general tendency to shrink them. Plastics and elastomers under bending are susceptible toward stress corrosion cracking if attacked by esters. Hence, stress-corrosion cracking has to be considered explicitly in the case if esters are used. Since hydrocarbons, like PAO have a tendency to swell elastomers, the addition of esters may counteract such that the effect is neutralized. As a fact synthetic oils based on PAO are additivated by addition of 10 or 20 % esters per weight to create this effect.

Low Temperature (Pour Point) Properties

Di-Esters, e.g. DOA, DOS are very useful in temperature ranges that undergo -40°C. This effect might be explained by the lack of hydroxyl groups that might associate at low temperature via hydrogen bridging, but also as a consequence of the crystallization hindrance due to the spatial structure of esters which does not allow a dense crystal packing. In contrast, esters may be designed such that their low temperature properties are lost, just by changing their structure. Also, if the number of polar groups increase the tendency to molecular association increases, and hence the pour point rises.

High Temperature Properties

High temperature applications in the use of esters are achieved by
- Sterical hindrance of the ß-Position in the Alcohol
- Use of Aromatic Nuclei in the Ester structure

As a specialty esters may rearrange within their structure via a preferred six-membered cyclic intermediate that creates an alkene on one side and a carboxylic acid on the other side (Figure 32).

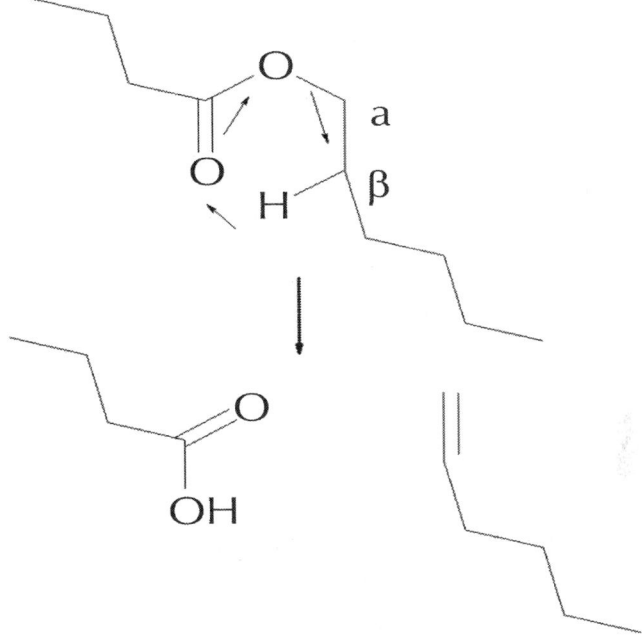

Figure 32: Decomposition of esters via cyclic rearrangement.

Degradation of esters via such mechanisms takes place at ambient temperatures, e.g. by copper activation even at 70°C. The formation of carboxylic acids and alkenes may lead to corrosion and unfavorable deposits on metals. In the case of blocking the ß-position, as in the NPG and TMP esters, the cyclic rearrangement is blocked and the ester does not undergo the thermal degradation. Such oils are commonly used as turbine oils.

Side Reactions

Hydrolysis

Ester Oils generally hydrolyze by interaction with water. The hydrolytic process is somehow the reverse reaction how esters form. The attack of water is enhanced if alkalinity is present but also acids may catalyze the hydrolysis. Common understanding states the attack of so called nucleophiles, like water at the carbonyl C-atom, followed by rearrangement sequences, leading to carboxylic acid and alcohols. (Figure 33)

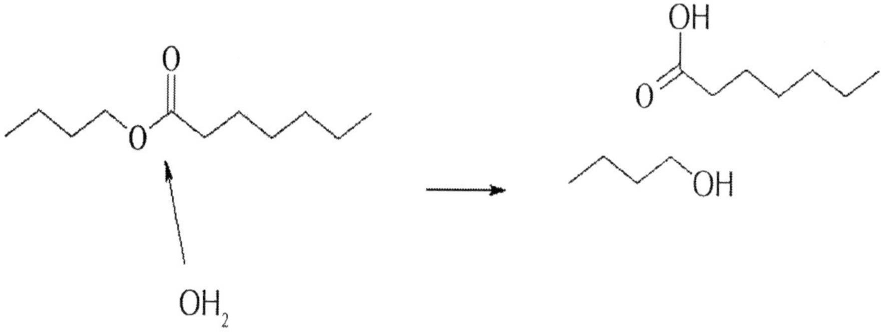

Figure 33: water-based cleavage of Esters toward carboxylic acids and alcohols.

Catalytic hydrolysis of ester oils also take place at metal surfaces, e.g. under tribological conditions. Formation of carboxylic acids may lead to corrosion as a consequence.

Biodegradation

Esters may decline under the interaction of bacteria and combust. Biodegradation is observed in the case of vegetable oils, e.g. glycerol esters, seldom on technical esters. In principal biodegradation cleaves esters, like water does to carboxylic acids. Biodegradation as a complex process does not stop there but lead to further products. Esters may oxidize as described in mineral oils and PAO at the organic tail. As a specialty they may undergo hydroxylation at a side position followed by trans-esterification to lactones. The lactone sequence is described already in the mineral oil section. Lactones are observed if esters, but also PAO are decomposed on iron at higher temperature. Infrared Spectra show absorption at 1800 -1760 cm^{-1} caused by lactone formation (see also chapter of antioxidants). (Figure 34)

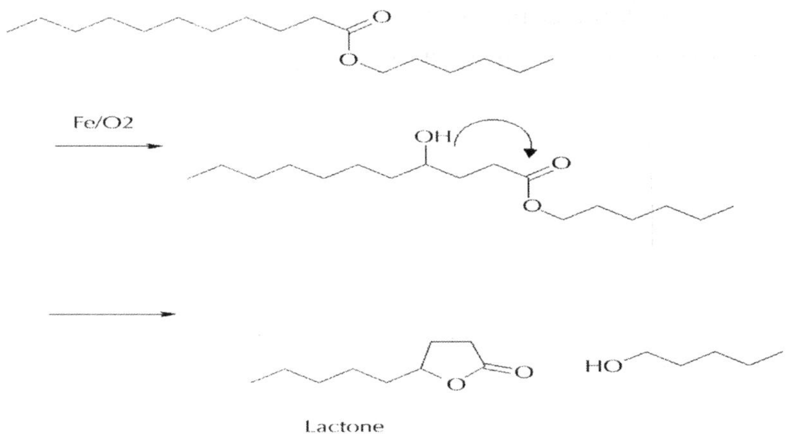

Figure 34: Lactone formation by side-chain oxidation of esters.

Other Esters

Esters may be created, as already mentioned by reaction of acids, in a different way as carboxylic ones. Prominent representatives are esters derived from phosphoric acid. Phosphorous offers two main oxidation states (+III and +V) from which acids are derived. Depending on the oxidation state and the alcohols, phosphoric esters are different in use. Also phosphorous overtakes the role of anti-wear activity in such substances. [9]

A common representative is Trilaurylphosphite (Figure 35).

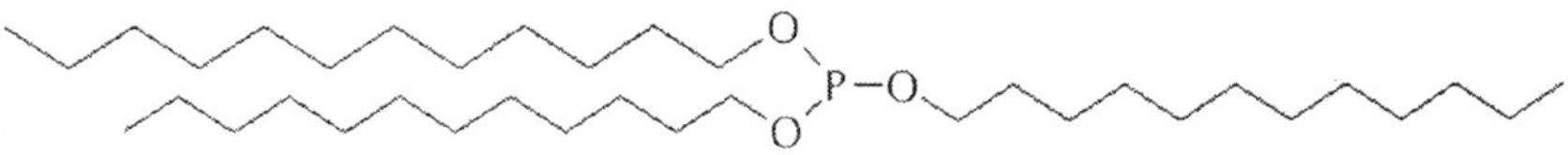

Figure 35: Trilaurylphosphite as a representative of Phosphinic Acid Esters.

Phosphoric Acid Esters

Phosphorus in the oxidation state (+V) creates a plenty of variant Acids, such as Orthophosphoric Acid, Diphosphoric Acid, Triphosphoric Acids switching into each other. (Figure 36)

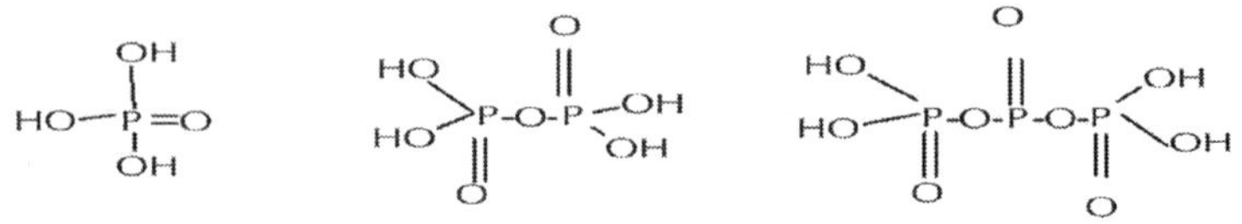

Figure 36: Representation of Phosphoric Acid Ester.

Phosphoric Acids are created by reaction of either phosphoric acid anhydride with alcohols or phosphoric acid derivatives, e.g. $POCl_3$ (Phosphorous-Oxi-Chloride) with alcohols. Aliphatic alcohols are in use, but also Phenols [10].

Formation of Phosphoric Acid Ester

Reaction of aliphatic alcohols, e.g. hexanole, with phosphorous pentoxide leads to hexylphosphate. In general some acidity remains due to insufficient esterifications. As a consequence those esters are often neutralized with amines to give amine phosphates. [10] (Figure 37)

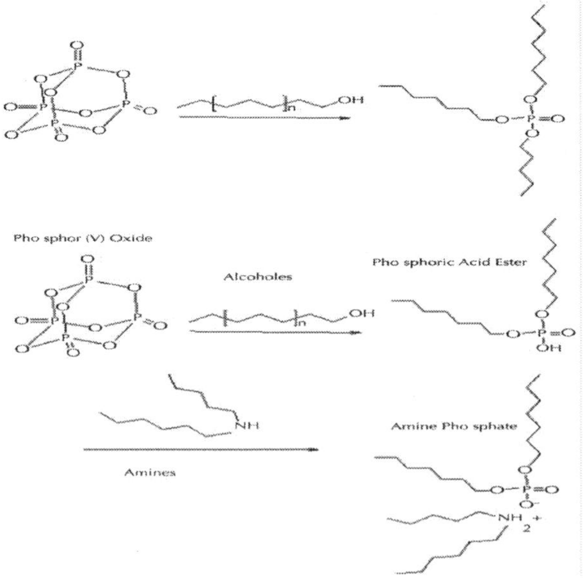

Figure 37: Phosphate Esters and Amine phosphates.

Amine phosphates are widely used as anti-wear and anti-corrosion additives in all kinds of applications. Phosphoric Acid Esters derived from Phenoles are shown below. (Figure 38).

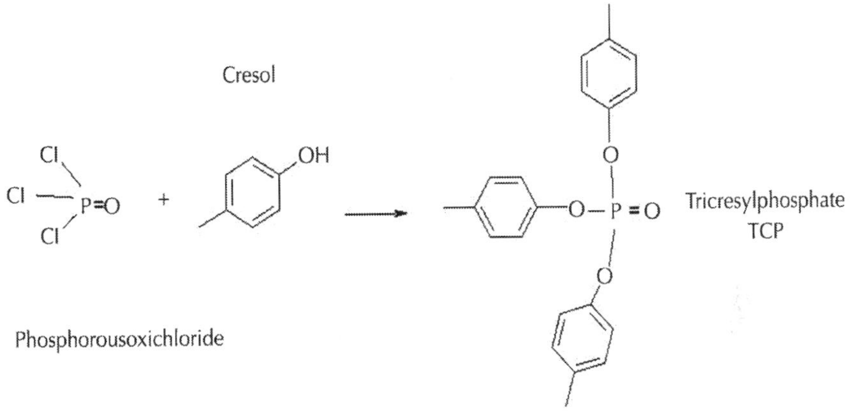

Figure 38: Arylphosphates derived from Phosphor Oxide Chloride Reactions.

In a different reaction Scheme Phosphoroxichloride reacts with alcohols. Those reactions are convenient to come to aryl phosphoric acid esters. Arylphosphates are somehow used to come to non-flammable high temperature lubricants at temperatures beyond 200°C.

Apart from the use as base oil phosphoric esters like Tricresylphosphates, based on the reaction from Phosphorous Oxichloride with Cresol (Methylphenols) are common additives for lubricants in bearing industry.

Whilst TCP with the methyl group in the para-position is seen as hazardous, TCP isomers in the ortho is registered to be highly toxic. Also mixtures of TCP isomers, due to the content of the highly toxic ortho isomer are registered as highly toxic. TCP, despite its superior behavior as AW additive for bearing lubrication is restricted for use (Figure 39).

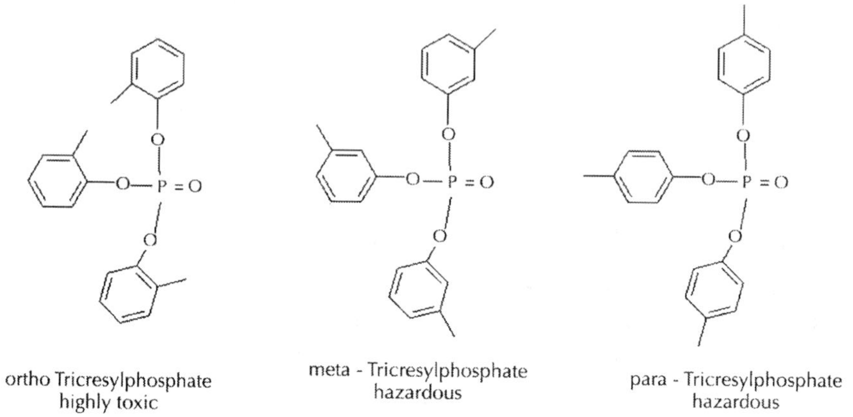

| ortho Tricresylphosphate highly toxic | meta - Tricresylphosphate hazardous | para - Tricresylphosphate hazardous |

Figure 39: TCP and some isomers.

Use of Thiophosphorylchloride as precursor, the reaction with Phenole leads to EP/AW additives like Triphenylphosphorothionate (TPPT) and its derivatives. (Figure 40)

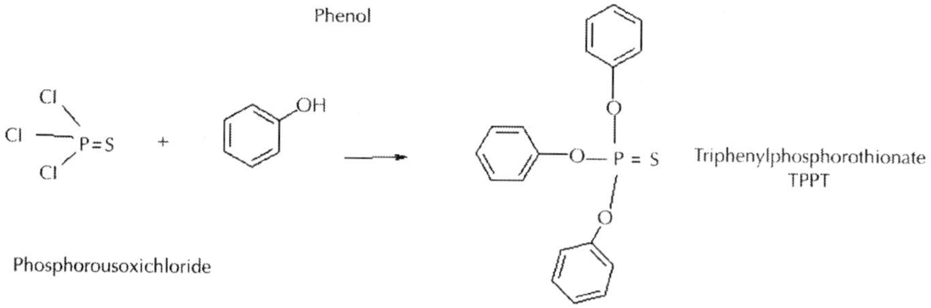

Figure 40: Synthesis of TPPT.

TPPT is widely used as nonmetal EP Additive as a substitute for Zn and Molybdenum Dithiophosphates. Due to its thermal stability, TPPT undergoes reactions at higher temperatures (>100°C). As to the fact that TPPT is ashless and starts to react at higher temperatures, it is a preferred additive in high temperature lubrication in combination with sterically hindered esters and PAO. In contrast to TCP, TPPT is not registered to be toxic, even more, the use of TPPT is allowed at level of 0.5% per weight for incidental food contact.

POLYGLYCOLES (PG)

General

Synthesis of Polyglycoles starts from Epoxides, obtained by catalytic oxidation of Alkenes from Petrol- or hydrocarbon chemistry. Polymerization catalysed either by acids or alkaline result in the formation of polyglycoles. In the case of alkaline catalyst, e.g. alkoxides on half of the PG contains a hydroxyl group while the end is capped by an ether function. (Figure 41) [1, 2, 7, 10]

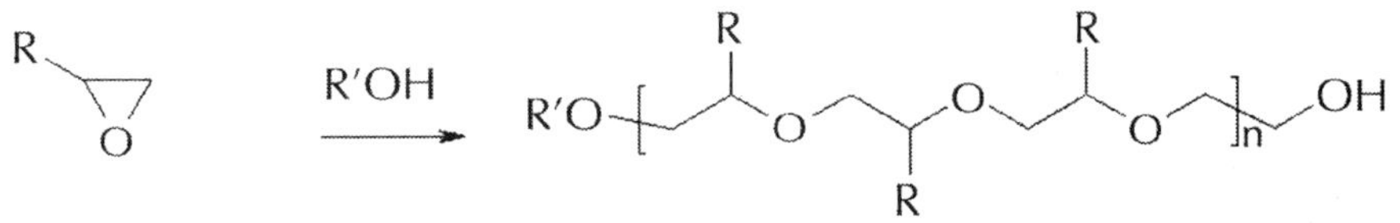

Figure 41: General Formation of Polyglycoles by alkaline catalytic polymerization of Alkene Epoxides.

The choice of either different alkenes (Group R) or alkoxides (R′) leads toward a huge variety of PG, all of them with different chemical and physical properties.

TABLE 1: PG and their data and applicability

Polyglycoles			PEG	PPG-PEG	PPG	MG
Description	Ethyleneoxide Polymer			Mixed Polymers Ethylene/Propylene Oxide	Proyleneoxide Polymers	Butyleneoxide Polymers
Chemical Data						
Physical Data	Density	aPProxx-	1	095-098	035-038	095-038
	Flashpoint	aPProx-				
	Pomp° int	aPProx-				
	Water miscible(%)		100	partially		
	Hydrocarbons		non miscible	partially	partially	partially
	Ester Oils		partially-full	partially-full	partially-full	partially-full
	Other PG	PPG	partially-full	partially-full	partially-full	partially-full

		PBC	partially-full	partially-full	partially-full	partially-full
Tribological Data	Viscosity, 40°C	Range	32 - 46	32- 100	32- 100	32- 100
	VT Coefficient	Range	180-_> 200	180-_> 200	180-_> 200	180-_> 200
	VP Coefficient	Range				
Others	Seals	PIER	compatible			
	ABS	ABS	compatible			
Paintings	Paintings		not compatible			

Technically only a couple of variances are produced in a larger scale, such as:

- *Polyethylenglycoles (PEG)* where Ethylene Oxide is the starter
- *Polypropyleneglycoles (PPG)* where Propylene Oxide is the starter
- *Polyethylene- Polypropylene Oxide Mixtures* started from mixtures or Ethylene and Propylene Oxide

Table 1 offers an overview across the most common PG their data and applicability.

Single addition of long chain alcohols lead to the formation of fatty alcohol ethoxilates, for use as non-ionic detergants and dispersants in lubricant formulations, as silicone free defoaming and emulsifiers for lubricant formulae.

In general PG are not thermally stable by themselves and tend to decompose by emission of volatile degradation products, e.g. low boiling compounds, such as aldehydes, ketones, acids and others. Due to this behavior PG are used in high temperature applications where the formation of polymers and lacquers due to heat induced degradation of lubricants is not convenient, for example high temperature chain lubrication.

Presence of alkalines, such as overbased sulphonates, widely used in motor oils, as corrosion inhibitor lead to multiple cross-reactions with the decomposition products of PG (aldol reactions): Results of the aldole reaction are tars, sludge and slurries in the system. In consequence corrosion resistance of PG should always be carried out by acidic corrosion inhibitors, such as succinic-esters, Zinc-Naphtenates or Phosphoric partial esters. (Figure 42)

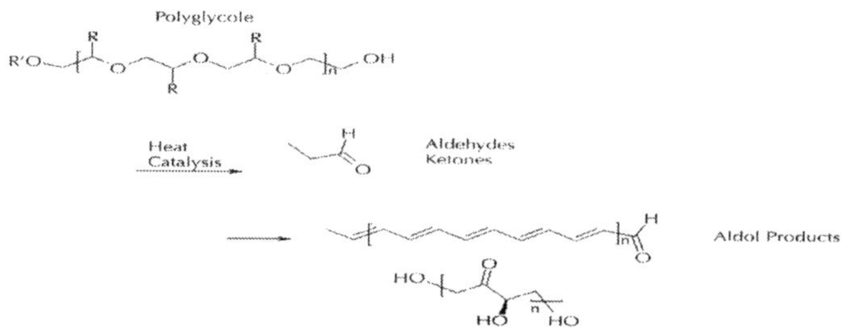

Figure 42: Aldole sludge formation in PG by use of alkaline.

It has to be considered that PG are poorly soluble even amongst themselves and should be carefully checked. In general their solubility in mineral oils is poor, better in esters (depending on the structure). However, PG needs to be stabilized by antioxidants in order to prevent the early thermal degradation. By doing so, the application of PG is enhanced significantly, such, that even applications temperatures > 160°Care approached.

Convenient stabilizers are Phenyl- -Naphtylamine, Phenothiazines or Alkyldiphenylamines. The amount should be adapted to the application.

In general PG offer very high viscosity indices, mainly above 160 (compared to mineraloils at ranges from 20 (alkylnaphtalenes), napthenics (70), paraffine base solvates (110), Poly- -Olefines (140).

This high VI allows reducing the calculated viscosity in a given tribological application down to one or two levels. For example, if in a given application ISO VG 320 (320 mm^2/s, at 40°C) is calculated for a mineral oil with a VI of 100, this viscosity maybe reduced by use of a PG down to 220 mm^2/s or even 150 mm^2/s. Pour points are low in the case of PG, very often in the range of -30.. – 40 °C even. Reaching the Pour point, PG tend to form highly viscous liquid, however, crystallization-inhibited. As a fact of this huge increase, PG is not for use even at temperatures above the pour point. Realistically PG is not suitable in the vicinity of their pour point. Therefore they are not very good low temperature base oils compared, for instance, with esters or PAO fluids. Due to their chemical structure PG are somehow strong solvents, e.g. paintings. In the case PG is used, the system has to be checked whether the paintings of the tank, the machine housing or others are affected. Dissolved painting from the tank may cause severe problems in the oil circuit by blocking filters. Additive response, known from standard applications, may change seriously by use of PG due to their different solvent capability. Extreme Pressure Additives have to be checked in their performance if used in PG. Normally anti-wear and anti-friction additives may be decreased in their content.

As a fact of the presence of epoxides in PG and due to their cancerogenic potential, the use of PG formulations drops down.

Polyethylene Glycols (PEG)

Polyethylene Glycols are made from ethylene oxide by polymerization (Figure 43).

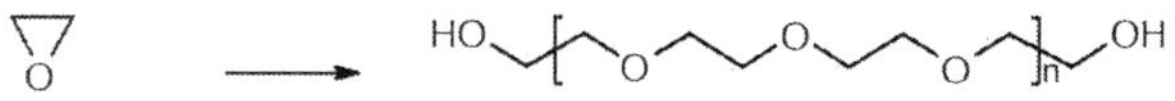

Figure 43: PEG formation and structure.

PEG, apart from its wide use in cosmetic industry is completely water miscible. Due to its water miscibility PEG is only or sparingly soluble in hydrocarbons. Compatibility of the PEG with a given fluid has to be checked before use. PEG, as facts of its water miscibility will uptake water without separation. In case of the use of PEG in applications within water environment the water ingress should be checked carefully. Effects of water ingress are increasing threat of corrosion and thinning due to the mixture.

Use Of Peg

Water miscibility is of use in non-flammable hydraulics in coal mining industries, but also in applications of pharmacy and food processing. In general PEG is allowed within the FDA regulation to be safe for incidental food contact. Due to the positive effect of sliding especially in worm gears PEG is somehow recommended for use in such applications. [2, 6, 7, 9]

Polypropylene Glycoles (PPG)

Polypropylene Glycols are made from Proplyene Oxide by polymerization by use of butoxides leading to a half ether structure (Figure 44):

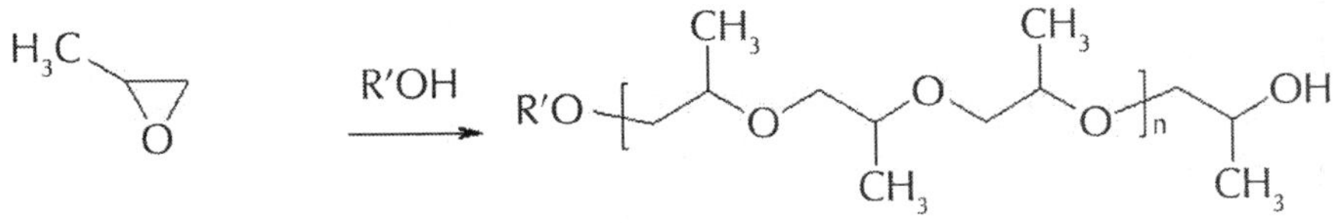

Figure 44: PPG Structure.

Due to the additional methyl group in the structure water miscibility drops down (contrast to PEG) while the oil miscibility promotes. Also by choosing longer alkyl chain butoxides, PPG structures may be obtained with enhanced oil solubility.

While PEG is highly dissolved in water, PPG forms droplets immersed in the water. Due to this fact water separation out of PPG is difficult to achieve.

The partial solubility and immersion of PPG in water causes a very high fish toxicity. PPG should never be used in the case of its break out-in free lands or water

Use Of PPG

PPG is commonly used as high temperature circuit oil, e.g. calandars, compressors, high temperature chain lubrication. All over PPG has to be stabilized by acid corrosion inhibitors, e.g. phosphoric partial esters and antioxidants like Phenyl- -Naphtylamine [2] [3] [5] [8].

Polybutylene Glycoles (PBG)

PBG are seldom in use and made consequentially from Butylene Oxide polymerization by use of alkoxides, leading to half esters (Figure 45)

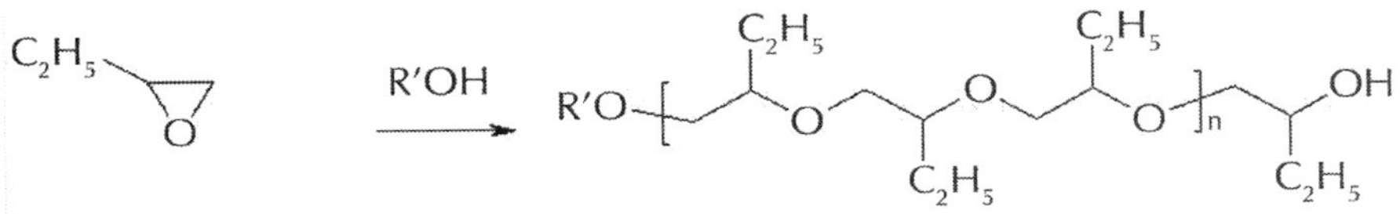

Figure 45: PBG Structure.

Use of PBG

PBG is useful for enhancing the solubility of additives, boosting the viscosity index of mineral oil variants.

Alcohole Ethoxilates

Alcohole Ethoxilates are formed by a cross reaction of epoxides with alcohols. (Figure 46)

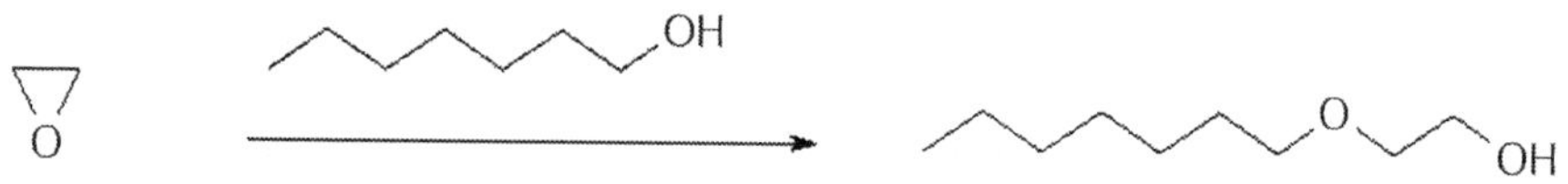

Figure 46: Alcohole Ethoxilates.

    The use of long chain alcohols leads to alcohole ethoxilates being used as non-ionic surfactants, emulsifiers and dispersants in multiple applications, e.g. hydraulic oils with dispersant capability, cutting fluids, and dispersants for applications where sludge is expected.

    Due to their non ionic nature alcohole ethoxilates are widely compatible in lubricant formulations.

SILOXANES

Siloxanes (Silicone Oils) are common lubricants in multiple applications, such as food and pharmacy, but also in applications where special low friction properties are demanded [2, 3, 8]

    In general silicones are the result of alkylchlorosilane hydrolyses [10] (Figure 47).

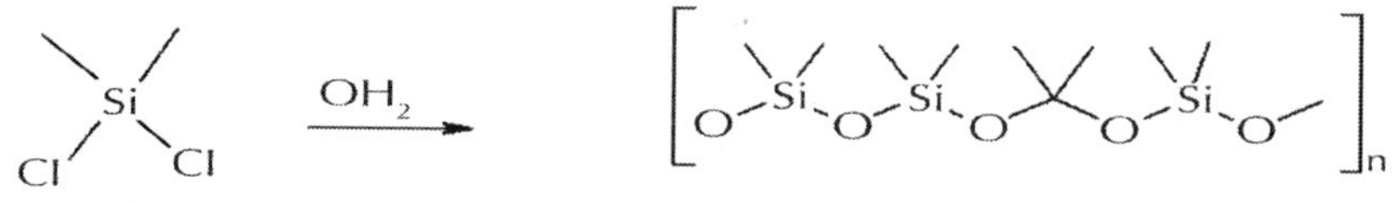

Figure 47: Scheme of Silicone Oil formation.

Side groups are methyl, but also phenyl groups leading to polydimethylsiloxanes or polyarylsiloxanes. Mixtures of methyl and arylsiloxanes are in use with different spreading between in the side chain (Figure 48).

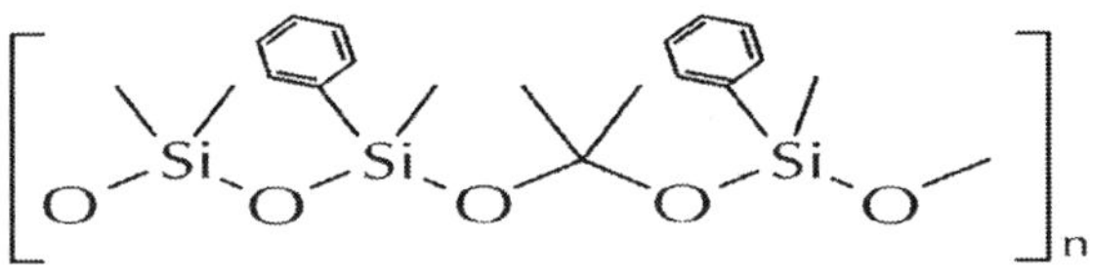

Figure 48: Polymethyl-Aryl Siloxanes.

Siloxanes with different structures are not generally miscible amongst each other. Miscibility has to be checked carefully. As mentioned, siloxanes are widely used in lubrication technology due their exceptional properties concerning low friction capability, high temperature stability and low toxicity in various applications. Prominent applications are starter components in cars, valves in food industry, slow speed bearings and high temperature applications where arylsiloxanes are in use. Siloxanes creep widely across surfaces and may cause problems in coatings, lacquering and paintings.

POLYFLUORINATED POLYETHER (PFPE) BASE OIL

PFPE Base Oil is created by polymerization of Perfluoroepoxids. Structure of PFPE is similar to polyglycoles but with overall substitution of hydrogen by fluorine [2] [3] (Figure 49)

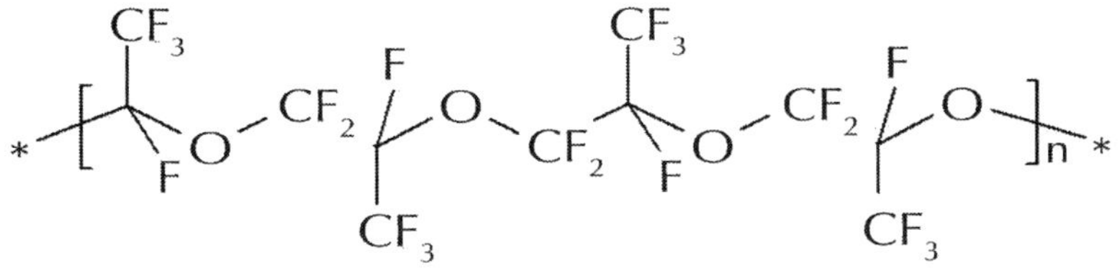

Figure 49: PFPE Base Oil.

Due to the effective shielding of the C-O-C backbone in the structure of PFPE by the trifluoromethyl side chain group PFPE are completely insoluble in water, inert toward alkaline and acids and even oxygen.

PFPE Base oil is used for high temperature purposes and in the presence of aggressive media, mentioned above in junction with PTFE thickener. (Figure 50)

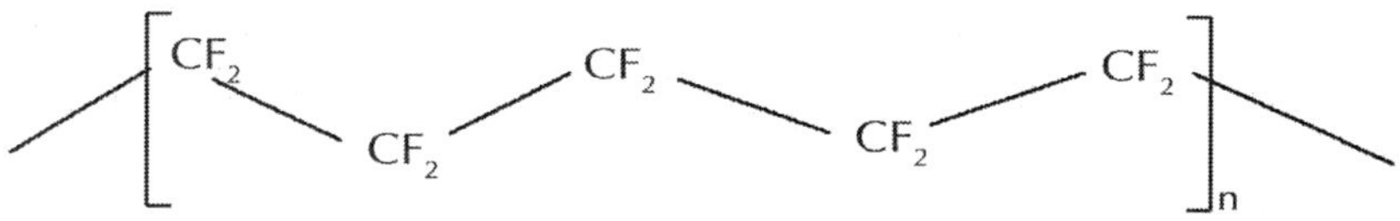

Figure 50: PTFE as thickener for PFPE.

PFPE sparingly adheres to metal surfaces due to droplet formation. The low adhesion causes creeping across surfaces and mal-lubrication if the surfaces are not cleaned thoroughly. Creeping and low adhesion may cause low friction in certain applications. PFPE is insoluble in most of the common base oils. Use of PFPE hence is restricted to the fluorine group of base oil.

Inertness of PFPE and PTFE make greases suitable for incidental food contact lubrication.

High temperature combustion of PFPE may cause the emission of hydrogen fluoride and fluoro phosgene which makes PFPE formulations somehow corrosive, especially on steel alloy compositions. Due to this fact, the high temperature corrosiveness should be carefully taken into account in the case of PFPE use.

ADDITIVES

Additives in lubricants enhance base oil functionalities. Additive technology is in broad scale based on organic chemistry syntheses. From their origin they are found by chance, less than by a real scientific approach. Nevertheless, literature about their reactions is innumerous from the very beginning. [2, 3,9]

Modern additive technology commenced in the early 20th century and has progressed continuously due to advanced organic chemistry syntheses. Upcoming modern spectrometry has been used to clarify their structures and their reaction at different metal sites. [2, 3, 9]

Beyond the basic and industrial reaction mechanism studies, mixtures of additives have been studied extensively by industry and science over the years. Such studies reveal the mechanisms of compatibility and incompatibility of additives acting together at a given application. [9]

For example a functional mismatch is caused by diverse demanding addressed to additives, e.g.: additives acting against corrosion may interfere with additives that have to prevent metal surfaces against fretting or welding.

Modern additive technology is inevitable to reach the "for-life" goal of modern technologies. As "for-life" might be understood in a different manner by users, additive packages are developed during the decades adapted to a given customer demanding. For example, the demanding to get automatic transmission gear oil performances is achieved by additive packages that may not fit for wind turbine or paper mill applications. Hence, additives and their mixtures have to be selected carefully for each purpose. [2, 3, 8, 9]

In general there are no rules up to now to predict additive performances at a given technical application. As a consequence formulations have to be tested in forecast extensively to assure its functionality. Such testing is addressed by international and national regulations.

Additives may cover a distinct structure-property relationship. Since there are no scientific rules declaring on how a chemical structure of an additive causes a function all variations in additives have to be validated by tests.

Additive technologies have been revised many times during their history, either due to a change in demanding or due to their toxicity. Toxicity is a severe problem in additive technology, since no one knows their real long term biological and ecological effects. [9]

Since the validation of those different chemical additive structures causes tremendous costs, it is a fact, that additive free technologies or additive technologies with marginal content level are favored as future solutions.

The following chapter addresses additive technologies concerning extreme pressure, anti-wear functions and also corrosion-protecting and antioxidants.

Extreme Pressure (EP) and Anti-Wear (Aw) Additives

General

Extreme Pressure (EP) and Anti-Wear(AW) Additives are functional chemicals in lubricants with the task to separate metal surfaces in the case of heavy loading and to improve their resistance toward wear in the case of oil film break in the contact [9].

Machinery elements that start to run or stop due to emergency show pronounced loading due to a lack of lubrication, e.g. the oil does not separate the metal surfaces and the protection of the oil film drops down. At that point EP and AW additives are supposed to jump into the arena by causing reaction layers preventing the metal from direct rupture or welding.

Their chemical structures are found by chance. For example observations during drilling and maching show that tools perform better if lubrication is carried out by use of sulfurized oils derived from vegetables, mixed and heated with sulfur.

Later on intense research the nature and reaction started including modern surface spectrometry techniques. The transformation of EP/AW additives as a function of the nature of the surfaces, their loading, contact geometry, temperature and their structure shows a clear picture of structure-activity relationship. Also additives perform as a function of their chemical structure, but also as a function of their solubility in base oil and as a function of other additives being present. In that sense, it is shown that additives either may prolong service life but are also capable to shrink life.

Sulfur Additives

Sulfur acts as a powerful extreme pressure additive. The high reactivity, especially toward copper makes it unlike to use sulfur as element in tribology.

Sulfur embedded in organic framework acts as a powerful Extreme Pressure additive. Choosing appropriate organic structures the activity toward copper drops down. However, using sulfurized additives copper deactivation should be present anyhow.

Sulfur is added either by reaction of reactive organic precursors like alkenes and their derivatives by heating up with the element, or by polymerization sequences with activated sulfur precursors such as di-sulfur dichloride. Doing so, all kinds of unsaturated specie gives reaction products leading to sulfurized specie. Prominent representatives are reaction products of Isobutene with Disulfur Dichloride, or reaction products with terpenes (Figure 51) but also unsaturated carboxylic acid esters, like rapseed oil:

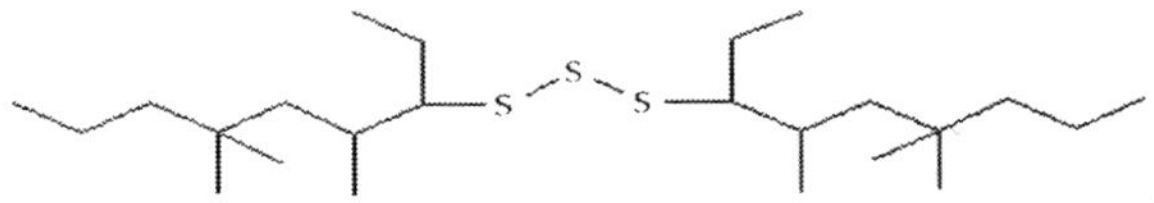

Figure 51: Didodecyltrisulfide as polysulfide representative.

Sulfurized Additives (S-Additives) are often used together with phosphoric acid esters, since the synergistic between those additives are known from the past. Doing so, gear oils may contain S-Additives with amine phosphate esters. Also extreme pressure additives containing Dialkyl-Thiophosphoricacidesters are prominent representatives in sulfur additive chemistry. (Figure 52).

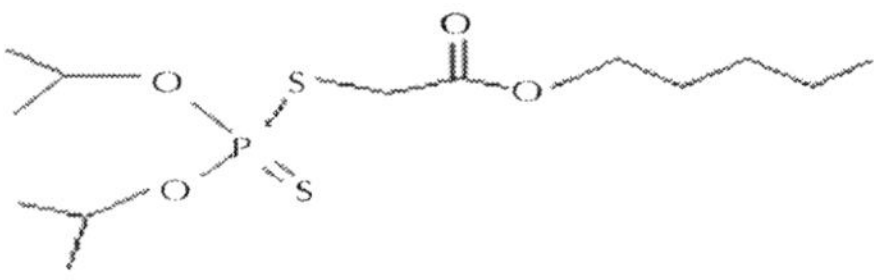

Figure 52: Thiophosphoric acid ester.

Dithiophosphates

Zinc- and Molybdenum dithio phosphates (ZndtP- ModtP)

Zincdithiophosphate (ZndtP) represent a prominent group of EP/ AW additives. They derive from the neutralization of Thiophosphoric Acids, obtained by ring opening of Phosphorous pentasulfide with alcohols, with Zn-Carbonate or Hydroxides. As a fact, the ZndtP differ strongly by their carbon-chain length. A couple of variants are achieved by choosing different alcohols in the ring opening sequence of Phosphorous (V) sulfide. From the structural perspective, ZndtP may be regarded as chelate complexes rather than a salt (Figure 53).

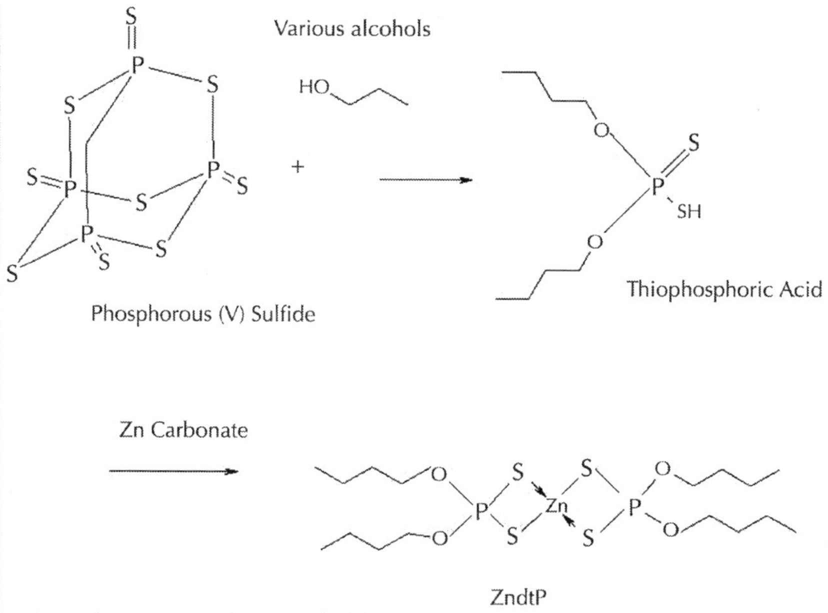

Figure 53: ZndtP from neutralization of Thiophosphoric Acid with Zn Carbonate.

Molybdenumdithiophosphates contains a Molybdenum [μ-oxo] Core, distinct compared to ZndtP (Figure 54).

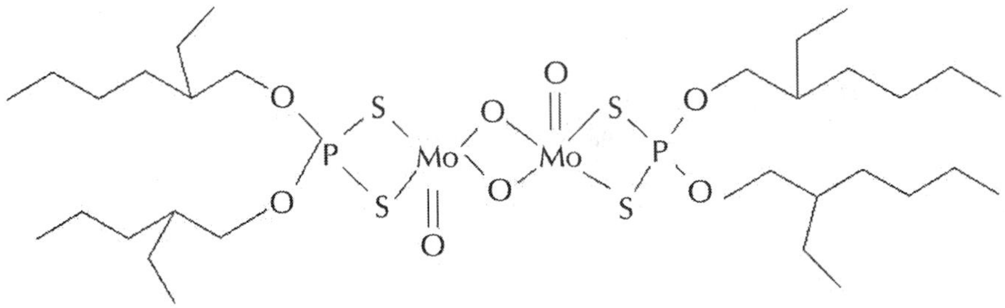

Figure 54: Molybdenumdithiophosphate.

Dithiocarbamates

Similar to Dithiophosphates, Chelat Complexes from Zinc, Molybdenum but also Bismuth and others may be formed by reaction of Thiocarbamic Acid with the metal precursors. Dithiocarbamic Acid is synthesized via addition of amines to Carbondisulfide. By varying the chain length of the amine different dithiocarbamates are achieved (Figure 55 and Figure 56):

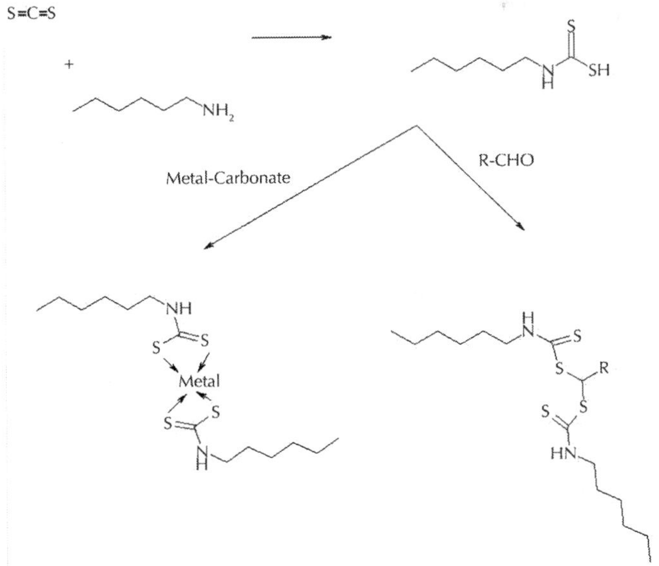

Figure 55: Syntheses of dithiocarbamates.

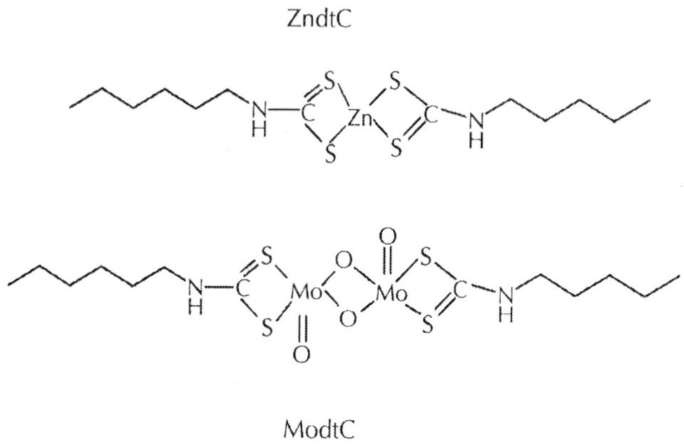

ZndtC

ModtC

FIGURE 56: Zincdithiocarbamate (ZndtC) and Molybdenumdithiocarbamate (ModtC).

Corrosion Protection

General

Within this chapter only iron as a chief element in technical application is considered.

Generally metal surfaces tend to corrosion if water, oxygen and probably salts, like sodium chloride are present. Corrosion may take place either by cathodic reduction of oxygen or by anodic oxidation of the metal. Charges, either positive (anode) or negative (cathode) pass the surface layer. [2, 3, 9]

Charge transport from the metal toward the outer region is hindered by the surface potential (over potential). Thus, corrosion processes have to overcome this potential and start after a certain induction period. Once, if this potential has been overcome the corrosion starts without hindrances by successive material transport. Materials transport ends up in a drastic change of the surface, mainly accompanied by a loss.

For iron as metal, the transport of the metal ends up in a flaky layer (Rust) that permits water and oxygen to penetrate. Due to this effect the rust process ends up in a total damage of the metal, especially in an environment that boosts corrosive processes.

Counteracting corrosion, the initial processes of charge transportation have to be blocked. Doing so, the over potential, e.g. the natural barrier of charges passing the surface has to be increased by creating additional layers on the metal surface (Passivation) or by creation of stable, insoluble complexes, formed by interaction of the surface atoms with a complex builder.

Passivation of iron surfaces and enhancing the over potential is achieved by deposition of chromium layers that cause a thin and gas-dense closed layer on the metal. Thus, chromium is a powerful inhibitor toward corrosion processes. As the charge transport phenomena occurring on iron surfaces are cathodic or anodic and vice versa, this process could also be stopped by offering an anodic victim like a zinc coating.

Additives that create a corrosion protection are in general dissolved in a carrier base-oil that spreads over the surface. Due to their adapted functional groups a physical binding toward the surface starts to create a layer. In order to create an appropriate corrosion protection this layer has to be packed dense to avoid the penetration of water and oxygen. This is realized by strong dipolar groups and oil soluble tails with a marginal demand in lateral spacing, e.g. long, - unbranched alkyl chains.

Else, passivation also is achieved by placing insoluble complex builders onto the iron surface, like phosphates are. Iron phosphate builds up a close dense insoluble layer on the surface.

Restriction of iron phosphate is indicated by the fact that, under certain conditions, phosphates start to get reduced forming posphanes. Phosph4anes strongly affect metals due to segregation of phosphorous at grain boundaries and releasing hydrogen into the metal. Hydrogen is detrimental to the microstructure by inducing, e.g. hydrogen enhanced local plasticit (HELP) or hydrogen induced cracking (HIC). Presence of phosphanes by reduction of phosphates takes place in acidic and reducing environment, e.g. presence of hydrogen sulfide, chlorides and others.

The following chapter will show some of the most prominent representatives of corrosion protectors.

Sulfonate-Chemistry

General

Sulfonates derive from sulfonic acids by neutralization with alkali, earth alkali –metals but also with metals from the transition group, for example zinc. Principally each sulfonic acid may be neutralized. In technical applications mainly alkyl benze sulfonic acids and dodecylsulfonic acid are neutralized. Production starts from alkenes out of petrol chemistry by addition sulfuric acid or SO_3.

Neutralization with either sodiumhydroxide, Calciumcarbonate, Magnesiumcarbonate or Bariumcarbonate leads to sulfonates: A = Sodiumsulfonate, B = Calciumsulfonate and with excess Carbonate to over based Calciumsulfonate (B'), Magnesiumsulfonate (C) and Bariumsulfonate (D) (Figure 57).

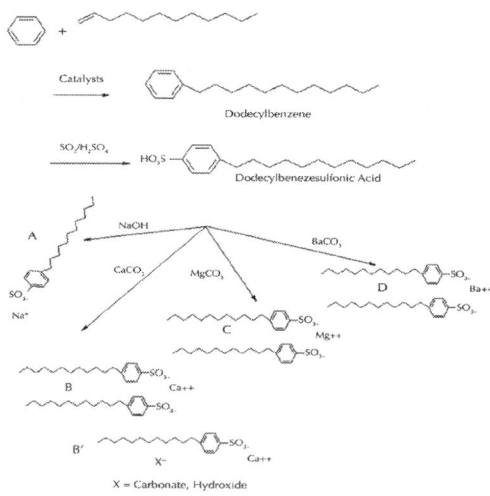

Figure 57: Sulfonic Acids and their Salts.

Carboxylic Acids And Derivatives

Carboxylic Acid and their derivatives, e.g. esters may act as metal corrosion protectors. While carboxylic acids are supposed to cause corrosion, some of them prevent. Rust preventing carboxylic acids are derivatives from -Aminoacids, like N-Oleylglycine. (Figure 58)

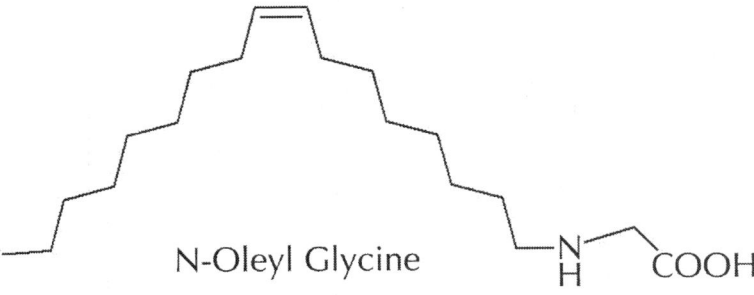

N-Oleyl Glycine

Figure 58: N-Oleylglycine.

N-Oleylgylcine acts as powerful emulsifier, even at low dosage. Rust protecting is due to the spread of water in the formulation over a big volume. N-Oleylglycine, even at low percentages also counteracts with EP/AW additives, driving their activity down.

Carboxylic Acids, derived from Phenoles such as Nonyl-phenoxiaceticacid is a non emulsifying corrosion protector but under prohibition, due to its irritating effects (Figure 59).

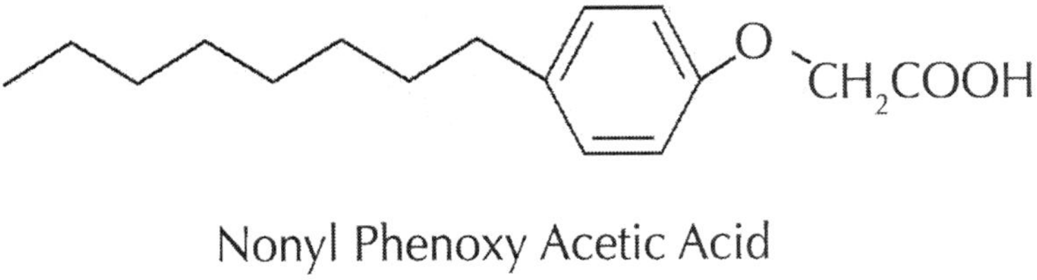

Nonyl Phenoxy Acetic Acid

Figure 59: Nonylphenoxyaceticacid.

Succinic Acid Derivatives, such as Succinic Half Ester of Octanole are powerful metal protectors, but also strong counteracting with EP/AW additives. Synthesis is carried out by reacting succinic acid anhydride with alcoholes (Figure 60).

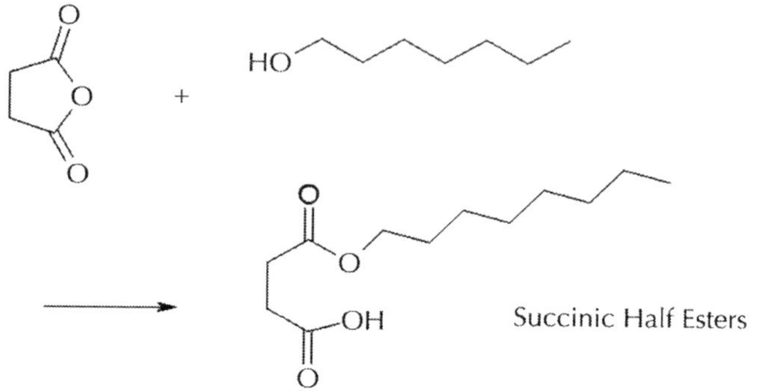

Figure 60: Succinic Half Ester.

Carboxylates, derived from neutralizing carboxylic acids with transition metals like Zinc, Lead, Bismuth lead to corrosion protection. Common acids are Napthenic acids or medium chain carboxylic acids like octanoic acid (Figure 61).

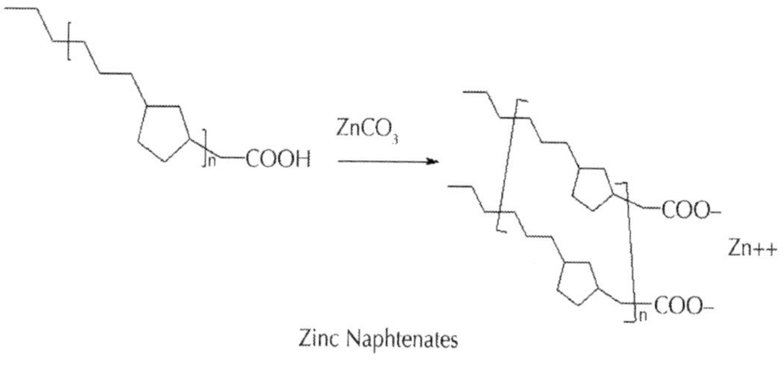

Figure 61: Zn (Bi) Carboxylates (Napthenate and Octoate).

Amine Phosphate Esters

Amine Phosphate Esters may act as anti-corrosion additives in addition to their anti-wear properties. Due to their synergistic properties and due to the fact, that certain amine phosphates are allowed as additives for incidental food contact, they are often found in all kind of lubricants (Figure 62).

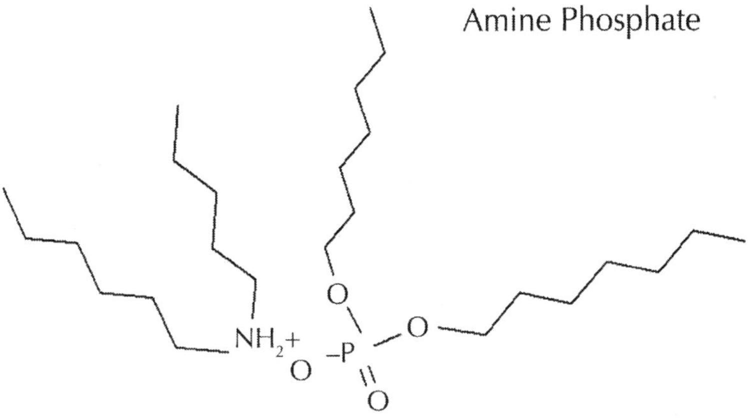

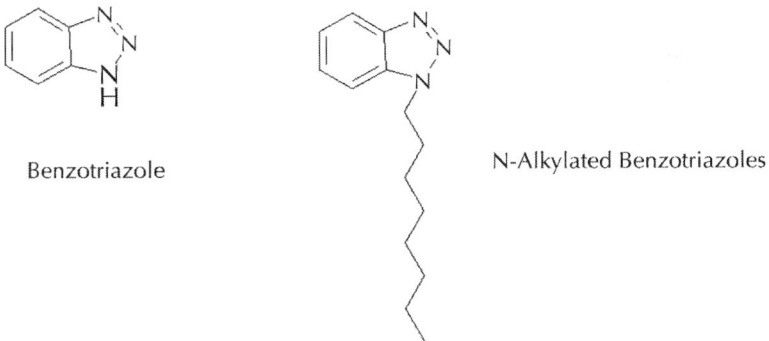

Amine Phosphate

Figure 62: Amine Phosphate Structure.

Amine Phosphates are powerful activators of copper and zinc and cause leaching of those metals from brass cages in bearings. Adding Amine Phosphates copper deactivators like benzotriazoles have to be present (Figure 63).

Benzotriazole

N-Alkylated Benzotriazoles

Figure 63: Benzotriazole and N-Alkylbenzotriazoles as Cooper Passivators.

Antioxidants (AO)

AO prevent lubricants from oxygen attack. Oxygen is, by nature, a diradical that undergoes several transitions. Electron uptake from metal surfaces by a cathodic transfer, leads to varieties of activated oxygen specie, powerful attacking hydrocarbon sites by abstraction of hydrogen, leading to peroxides, and carbon radicals. The carbon radical itself starts to stabilize by abstraction of hydrogen leaving an alkene as new product [10]. (Figure 64)

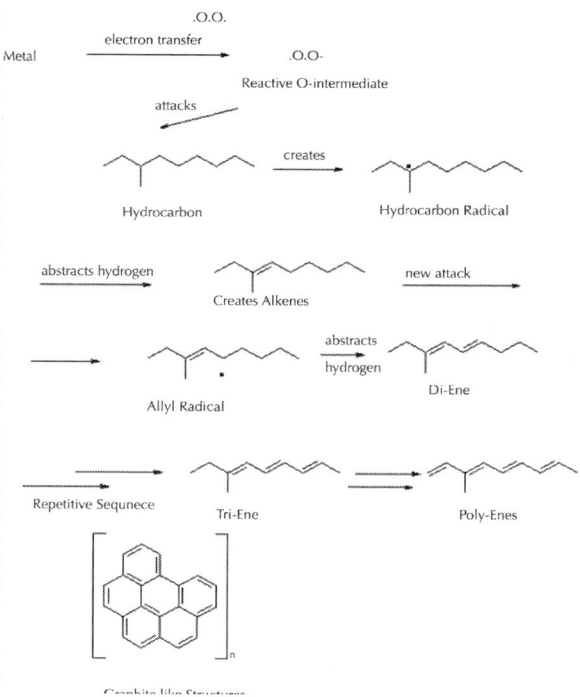

Figure 64: Oxygen – Hydrocarbon Attack sequence.

Due to radical stabilization the new formed alkene starts to continue the oxidation by sequential abstraction of hydrogen, forming di-, tri- and polyalkenes, but also benzene rings. Apart from the hydrogen abstraction, also oxidation takes place by attacking carbon radicals by oxygen. At least the products created by such this procedures are carbonyl compounds, e.g. alcohols, ketones, aldehydes, carboxylic acids and sometimes esters. PAO oxidation at metal surfaces, e.g. iron beyond 120°C results in the formation of lactones (esters that come up by internal reaction between an alcohol group and terminal carboxylic group) (Figure 65).

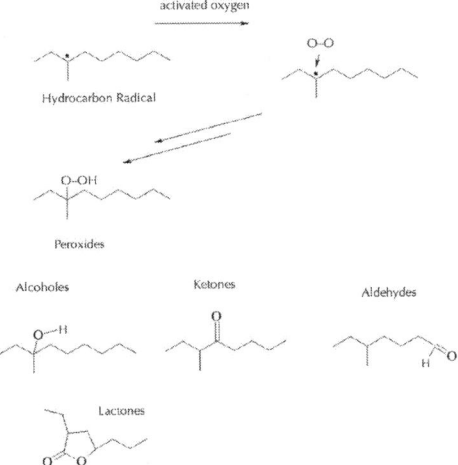

Figure 65: Oxidation sequence of Hydrocarbons toward carbonyl compounds.

Hence oxidation sequences dramatically change the original hydrocarbon chain. If once started it is self-accelerating till new, different and stable products are reached. Oxidation is unselective and takes place everywhere in the chain. Hence, plenty of products are formed by radical oxygen assisted processes.

Antioxidants in general prevent the base oil, quenching the oxygen attack by formation of stable radicals. Stabilization of the radicals is realized by a delocalization of the persistent AO radical, created by oxygen attack due to the presence of aromats in the structure (Figure 66).

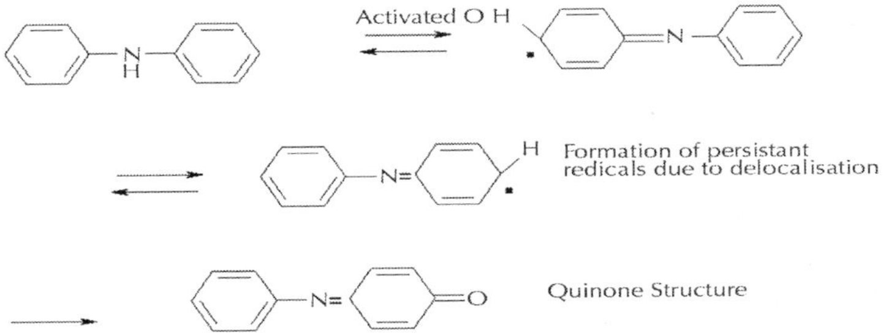

Figure 66: Principal delocalization of radicals created by oxygen attack.

The AO radical subsequently stabilizes to form new products like quinones. The quinone structure may form a dark colored charge transfer complex with the original antioxidant. Very often this causes strong discoloration of AO stabilized lubricants since the charge transfer complexes are very intense in color. Sometimes, for example in the case of polyurea greases, such charge transfer complexes may interfere with the grease structure in terms of solidification.

Persistent radicals formed by AO are dangerous in some cases. In the case of their accumulation in the system they are able to boost oxidation rather than to prevent. Dosage of AO hence should be carefully tested. Formation of either charge transfer complexes or oxidation products by the presence of AO may cause increased formation of sludge in the lubricant if the dosage balance is not appropriate.

Nearly all AO contain aromats as a base principle. Prominent AO candidates are Butylhydroxitoluene (BHT) (A), Alkyldiphenylamine), Phenyl- -Naphtylamine (PAN) (C) and various others (Figure 67).

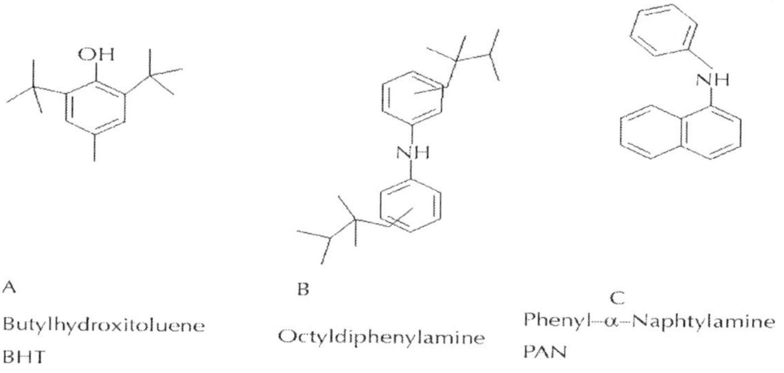

A	B	C
Butylhydroxitoluene	Octyldiphenylamine	Phenyl-α-Naphtylamine
BHT		PAN

Figure 67: Structures of AO: (A): BHT, (B) Alkyldiphenylamine, (C) PAN.

GREASES

General Remarks

Greases are defined apart from their chemical composition by the manufacturing processes. Thickener and oil, getting heated by stirring, start to dissolve. Getting cold, the process of stirring leads to a raw material where amorphous and crystalline structures are merged. The amount of crystals and amorphous materials depends on the nature of the raw materials on the one side and on the rate of heating and cooling on the other side. Rapid cooling causes homogeneous and amorphous structure, as particles are not able to grow to a large size. The raw grease, as effect of the mixture of solid structures has to be homogenized carefully. Homogenization leads to a smoothened appearance of the grease with a scale distribution of thickener particles as effect of the cooling process. Slow cooling generally leads to material with large sized particles as an effect of nucleation and crystal growth. Oil embedding in such structures is different due to the solid structure of the thickener. Stiffness and flowing capability may change as an effect of the merged structure. Greases, even in the case of identical chemical composition may differ significantly by their manufacturing process. Stiffness of greases is defined by the NLGI grade declaration, measured by penetration of standard cone into the grease. The deeper it's penetration the more liquid the grease will be. To get a constant value, the grease is worked by 60 strokes, then tempered to 25 °C and measured by cone penetration. NLGI grades are presented in table 2: [3]

TABLE 2: NLGI Grades of Greases

NLGI Grade	Cone Penetration in 1/10 mm
000	445 - 475
00	400 - 430
0	355 - 385
1	310 - 340
2	265 - 295
3	220 - 250
4	175 - 205
5	130 - 160
6	85 - 115

Oil Bleeding

Within grease the base oil is bound in different states. Some oil is weakly bound to the thickener nuclei and gets easy released. Oil, bound in micelles and large structures with van-der-Waals and dipolar bonding releases less. Oil release takes place due to centrifugal effects in speeding machinery elements, e.g. bearings, creeping across walls e.g. sealings enhanced by temperature. Successive loss of oil in grease may lead to its change in performance, accompanied by a malfunction. Oil bleeding is measured with different techniques. Within the most popular one the grease is sat on a sieve and pressed by a static load through it a given temperature. Bleeding is measured as a function of time. For bearings the long term bleeding rate should be less than 5 % per weight in 7 days. [3]

Dropping Point

Greases - if heated - start to get liquid at a certain point. Molten grease will leak out at sealing edges and may cause a malfunction of the grease. For bearings the thumb rule is given by dropping point minus 50 °C as the upper point of applicability. [3]

Soap Based Greases

Greases are soft solids, created by a thickener that gelates in suitable base oils. Gelling takes place by intense mixing of thickeners with the base oil, often accompanied by heating till the gelation is reached [3]. (Figure 68):

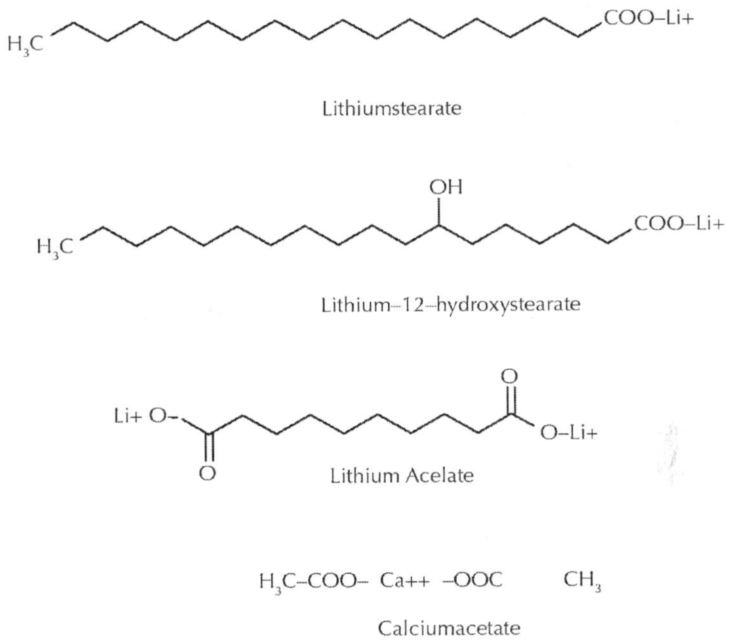

Lithiumstearate

Lithium–12–hydroxystearate

Lithium Acelate

$$H_3C-COO-\ Ca++\ -OOC \quad CH_3$$
Calciumacetate

Figure 68: Prominent representatives of thickeners for grease production.

Thickeners are all substances where gelling in the base oil is achievable. Prominent representatives are lithium and calcium salts of carboxylic acids, for example Lithium Stearate, Lithium-12-hydroxistearate, Calciumstearate, Calcium-12-hydroxistearate but also Calciumacetate. Lithium Complex Greases are created by the co-existence of lithium-12-hydroxistearate with dicarboxylic acids like Acelaic or sebacaic acid.

Calcium Complex Greases are composed by calcium acetate, Calcium Stearate and calcium-12-hydoxistearate as thickeners.

Salts of magnesium, barium and alumina are used for grease production but to minor extent.

Di And Polyurea Greases (Pu-Greases)

Urea Greases are often called PU-Greases in technical language.

Urea structures are realized by adding amines to isocyanates (Figure 69):

$$-N{=}C{=}O \quad + \quad -HN{-}R \quad \longrightarrow \quad -NH{-}CO{-}NH{-}R$$

Isocyanate   Amine   Urea

Figure 69: Urea Formation.

Di-Urea grease production take aromatic Isocyanates, like Diphenylmethane Isocyanate (Methylenbisdi-isocyanate, MDI) reacted with various aliphatic amines, like Cyclohexylamine, Alkylamines from C8 to C18 chain length.

Synthesis of the thickeners and grease formation is carried out simultaneously. Ester Oils, like trimellitic acid esters facilitate the synthesis by solving the precursors before the reaction takes place (Figure 70):

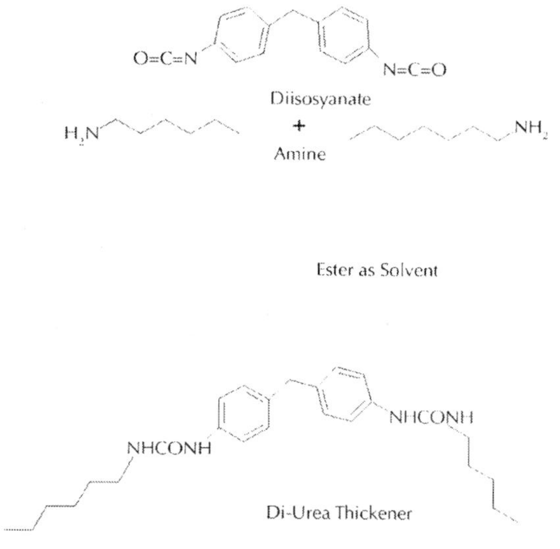

Figure 70: Formation of Di-Urea Grease.

Tetra- and polyurea Greases are created by mixing Di-Isocyanates like MDI or Toluenediisocyanates (TDI) with diamines, like ethylene diamine and monoamines, like Octadecylamine in suitable base oils (Figure 71):

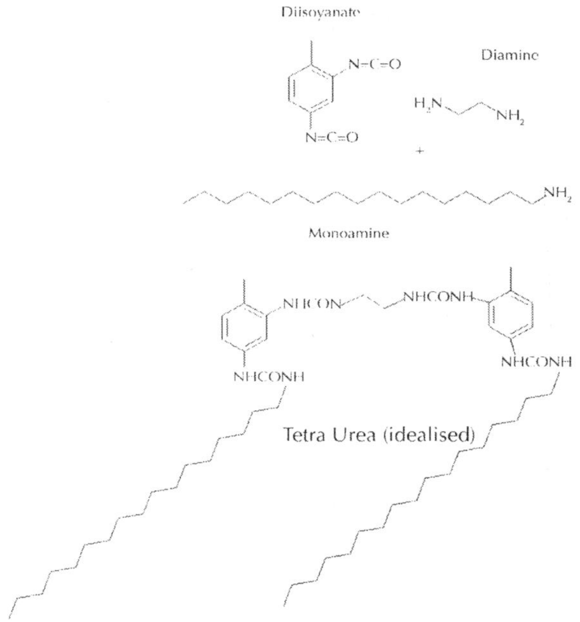

Diisoyanate

Diamine

Monoamine

Tetra Urea (idealised)

Figure 71: Formation of Tetra-and Polyurea Greases.

Urea Greases offer plenty nitrogen-hydrogen bridges within their structures. Concordant with the presence of temperature resistant aromatic nuclei and in junction with high temperature resistant base oils, they represent the group of high temperature grease "per se". As to the high variability of taking precursor amines, PU greases offer the possibility to adapt the grease to a given application, much more than soap greases do.

Polyurea Greases that start from tallow amine, tolyenediisocyanate and ethylene di-amine are in accordance with the US FDA regulations H1 (incidental food contact) if H1 base oil (like white mineral oil or PAO) is used. Also the modern EU REACH regulations are valid for polymeric structure.

As the polymeric degree increase, the thickeners may get insoluble and crystalline. Greases are no longer available due to this because a lack of gelling. Due to this fact, variances of PU Greases are restricted.

MDI and especially TDI are ought to be highly toxic by inhalation. Production of PU greases have to take care, than none of the precursors are free in air, nor present in the grease.

Some isocyanates tend to polymerize during production, rather than to react with the amine, especially at the end of the syntheses. Polymeric Isocyanates may remain in the grease and cause severe toxicitiy, especially if the greases are up -heated.

PU Greases are very sensitive toward ingress of OH – groups (e.g. alkalines, water, polyglycoles) as the nitrogen-hydrogen bridging is disturbed. Ingress of such pollutants may cause a change in consistency. Polyglycoles, if heated emit aldehydes that interfere with the NH groups in PU greases. This reaction may end up in making the solid PU liquid! PU greases thus should be monitored to those facts (Figure 72):

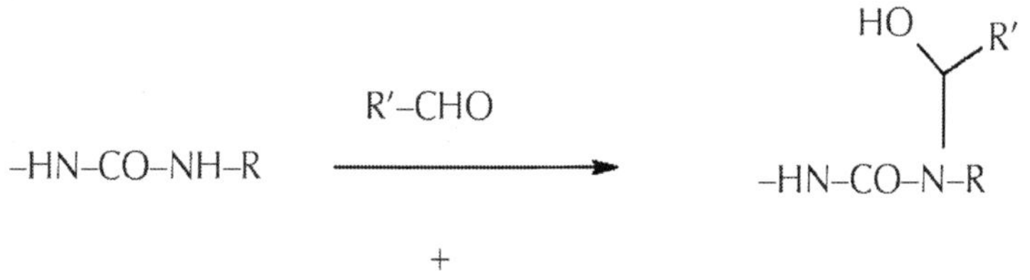

$$-HN-CO-NH-R \quad \xrightarrow{\quad R'-CHO \quad} \quad -HN-CO-N-R$$

Figure 72: Reaction of PU Grease and Aldehydes.

Other incompatibilities of PU Greases arise from mixtures with clay thickeners due to the presence of either OH (Si-OH) or NH functional groups if the clay is modified by organic amines.

Other Thickeners

Clay Greases-Structure and Use

Clay Thickeners derive from Alumina-Silicates. Due to their high surface and modification they are suitable for gelling base oils, e.g. Esters, Napthenic Base Oils, sometimes Silicones and Phosphoric Acid Esters. Clay Structure is generated by tetrahedral arrangement of Silica with insertion of alumina (see figure) in layers of approximately 1-2 nm distance. Water and other cations may be inserted in the space in between the two layers. Other cations, e.g. magnesium, may also be inserted in between. [3] (Figure 73).

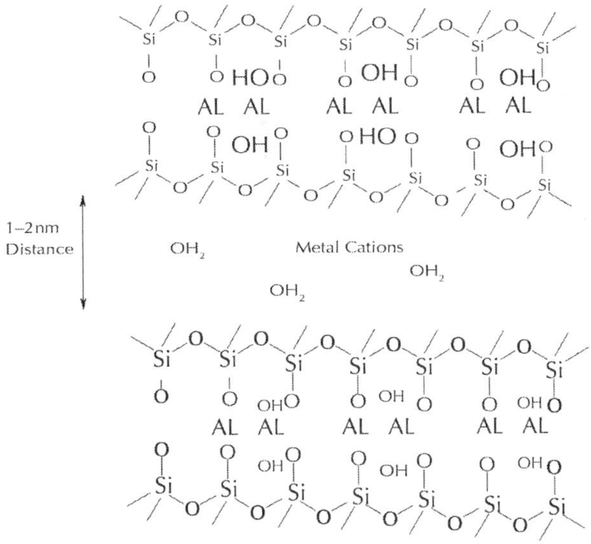

Figure 73: Estimated basic structure of clay.

Gelling takes place by adhesion and insertion of organic molecules in the structure, assisted by polar additives like propylene carbonate. Clay grease is produced by multiple milling the clay with appropriate base oil by addition of water suppliers like glycerol or Propylene carbonate at temperatures below 100°C. If water is lost the structure may break down during the manufacturing. Doing so, the grease produced is a buttery solid with no dropping point.

Use of Clay Greases

Clay greases are used for applications where the grease should not move out and for special high temperature applications, e.g. cement industry in slow motion bearings. Due to the inertness of the inorganic structure toward alkaline and acids, clay greases are preferred in applications where water, alkaline and acids are present, e.g. chain or bearing lubrication with such ingress. Clay is declared as safe for incidental food contact and allowed for lubricants in food industry (USDA H1 regulated) in junction with base oils like white mineral oil, PAO or esters that are allowed for this purpose.

Restrictions in the Use of Clay Greases

Restrictions for the use of clay greases are the presence of Lithium, - Calcium or Polyurea Greases that may interfere with the hydrogen bonding of the clay structure. Mixtures of clay and conventional greases should be evaluated very carefully. Clay greases are restricted in bearing lubrication strictly due to over rolling speed. In general the speed factor is limited to ndm (Average of outer and inner diameter of the bearing times the speed (revolution per minute)) of 100.000. Only slow moving bearings could bear clay lubrication.

SILICA

Silica is in use for thickeners as amorphous material, obtained by flame decomposition of Silica Tetrachloride (Figure 74):

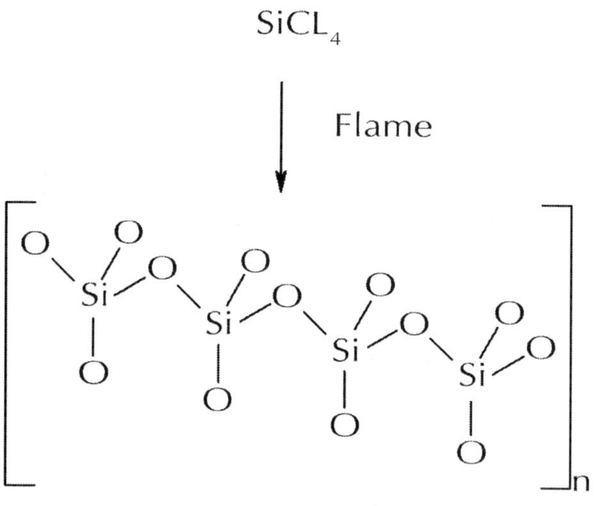

$$SiCL_4$$

Figure 74: Principal formation of amorphous SiO2 by flame combustion.

Silica, due to its powerful surface activity may be used as powerful thickener in low percentage for each kind of base oil. Greases obtained by mixing silica with base oils are transparent. The inorganic structure causes no dropping point for such greases. Silica Thickened greases cause steep and irreversible thickening by heating up due to the increase of internal hydrogen bonding. They never should be in use for high speed and high temperature rotating bearings, since they block their motion. The ndm (Average of Bearing Size times revolution per minute) is restricted to 100.000, hence slow motion. Due to the possible entrance of water, silica thickened grease is poorly water stable and should not be in use in applications where water (especially hot water) and alkalines are present. Alkalines react with silica to silicates, starting its degradation.

Polytetrafluoroethylene (PTFE)

PTFE is a convenient thickener in base oils for the purpose of incidental food contact, low friction properties and high temperature. The fluorine entity causes low activity toward oxygen. PTFE Grease is used in oxygen application (valves under oxygen impact), especially with PFPE.

CONCLUSIONS

Tribology is highly guided by physics and chemistry of the lubricants. Functionality of lubricants is given by their physics and their chemical structure. Modern understanding of lubrication hence allows the construction of lubricants appropriate to a given application to a certain extent. Under the conditions of full lubrication their physical properties, e.g. viscosity, viscosity-temperature and viscosity –pressure properties dominate over the chemical structure. Under such circumstances, the lubricant takes away heat (cooling function) from the mating contacts, but also wear and debris (cleaning function). Within a running – in period some reaction layers of lubricant constituents (additives) may be created. Basically those layers stay constant over time and do not change. On the other hand, if lubrication undermines the given roughness's of the mating partners, or overtakes the natural temperature limit given by the restrictions of organic chemistry (e.g. temperatures beyond 150°C), chemistry starts to perform reaction scenario highly related to the nature of the chemical structure of the ingredients in the lubricant. The basic reactions found here are radical reactions, as a fact of the presence of oxygen and iron. Within such radical reaction sequences hydrogen is abstracted, alkenes and alkynes are formed and their oxidation products (aldehydes, ketones, carboxylic acids and their derivatives). Additives, in general improve the lubricants by expanding their limits.

In general, lubrication fundamentals in tribology have overcome the alchemy of the past by numerous efforts taken by the scientific community.

REFERENCES

1.   Rudnik L.R., editor. Synthetics, Mineral Oils, and Bio-Based Lubricants. Boca Raton: CRC Press; 2005.

2.   Dresel W., Mang T., editors: Lubricants and Lubrication. 2nd Edition. Weinheim: Wiley-VCH; 2007.

3.   Klamann D. Schmierstoff und verwandte Produkte. Weinheim: VCH-Verlag; 1982.

4.   Mortier R.M., Fox M.F., Orszullik T.M., editors. Chemistry and Technology of Lubricants Dordrecht: Springer; 2010. http://link.springer.com/book/10.1007/978-1-4020-8662-5/page/1 (accessed 27 December 2012).

5.   Dowson D., Taylor C., Childs T., Dalmaz G. editors. Lubricants and Lubrication. In: Tribology Series 30 : Proceedings of the 21st Leeds-Lyon Symposium on Tribology. Amsterdam : Elsevier; 1995.

6.   Bloch, H.P., Practical Lubrication for Industrial Facilities. Lilburn: Fairmont Press; 2000.

7.   Stepina V., Vesely V. Lubricants and Special Fluids. Amsterdam: Elsevier; 1992.

8.   Lansdown A.R., Lubrication and lubricant selection: a practical guide. 3rd Edition. John Wiley & Sons; 2004.

9.   Rudnick L. R., editor. Lubricant Additives: Chemistry and Applications, 1st Edition. New York: Marcel Dekker, 2003.

10.   March, J., Advanced Organic Chemistry: Reactions, mechanisms, Structure. New York: Wiley-VCH; 1992.

Chapter 2

Lubrication and Lubricants

Nehal S. Ahmed[1] and Amal M. Nassar[1]

[1]Additives Lab., Department of Petroleum Applications, Egyptian Petroleum Research Institute, Nasr City, Cairo, Egypt

INTRODUCTION

Lubrication

The primary purpose of lubrication is to reduce wear and heat between contacting surfaces in relative motion. While wear and heat cannot be completely eliminated, they can be reduced to negligible or acceptable levels. Because heat and wear are associated with friction, both effects can be minimized by reducing the coefficient of friction between the contacting surfaces. Lubrication is also used to reduce oxidation and prevent rust; to provide insulation in transformer applications; to transmit mechanical power in hydraulic fluid power applications; and to seal against dust, dirt, and water.

The Lubrication Regimes

The modern period of lubrication began with the work of Osborne Reynolds (1842-1912). Reynolds's research was concerned with shafts rotating in bearings and cases this show in Fig.1. When a lubricant was applied to the shaft, Reynolds found that a rotating shaft pulled a converging wedge of lubricant between the shaft and the bearing. He also noted that as the shaft gained velocity, the liquid flowed between the two surfaces at a greater rate. This, because the lubricant is viscous, produces a liquid pressure in the lubricant wedge that is sufficient to keep the two surfaces separated. Under ideal conditions, Reynolds showed that this liquid pressure was great enough to prevent direct contact between the metal surfaces. Fig.2 taking a plain journal bearing as example, Fig.3 which is known as Stribeck curve summarizes the lubrication regimes by describing the relationship between speed, load, oil viscosity, oil film thickness, and friction.

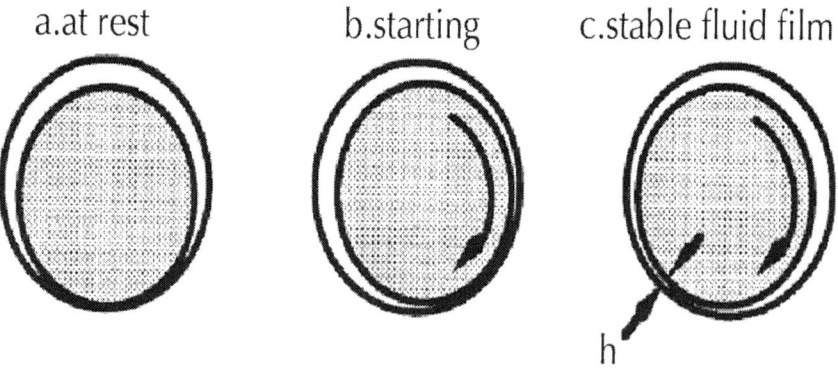

Figure 1: Three positions of shaft in a bearing.

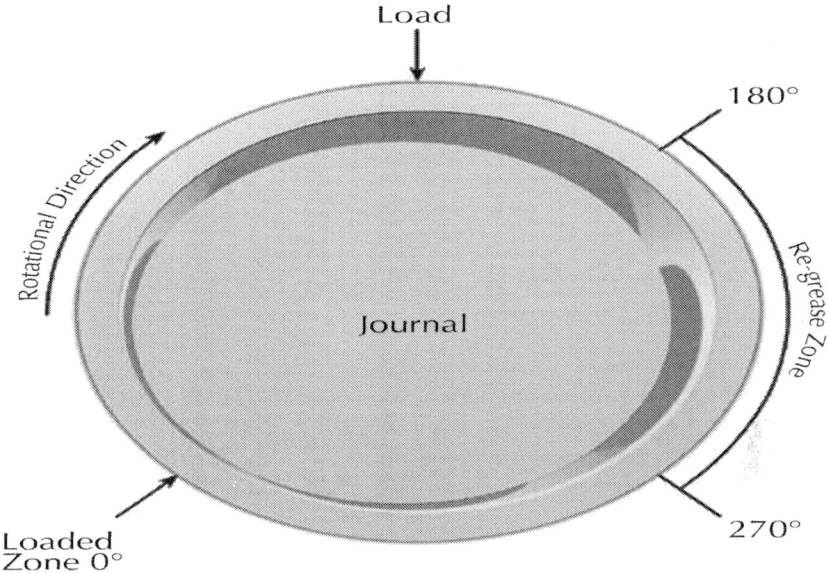

Figure 2: Plain Journal bearing.

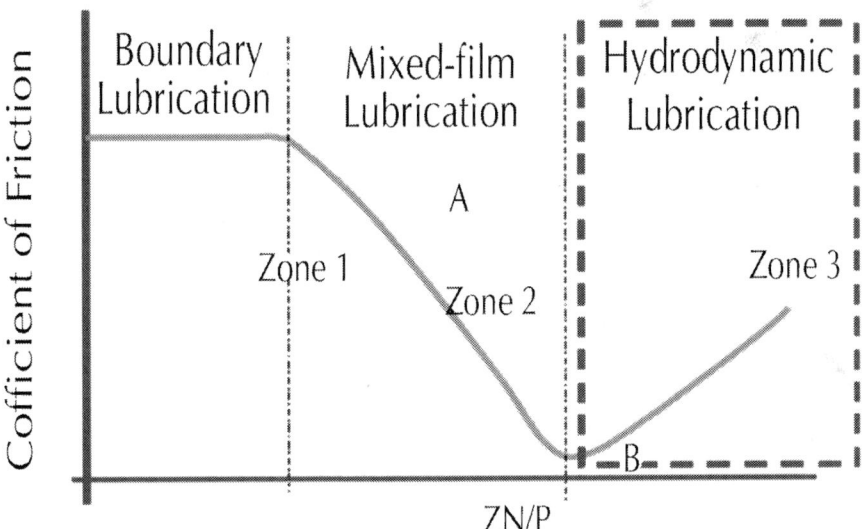

Figure 3: Stribeck curve.

In this graph, the coefficient of friction is plotted against the expression ZN/P (sometimes referred to as the Hersey number)

$$\text{Where} \quad ZN/P = \frac{oil\,viscosity \times shaft\,speed}{bearing\,pressure} \qquad (1)$$

As shown there are three distinct zones separated by points A and B. At B the oil film is just thick enough to ensure that there is no contact between asperities on the shaft and bearing surfaces. Smoother surfaces shift B to the left, while at point A the oil film thickness reduces virtually to nil. Zone 2, between A and B is known as the zone of mixed lubrication. Mixed-film lubrication is unstable at which increase in lubrication temperature causes further increases in lubrication temperature.

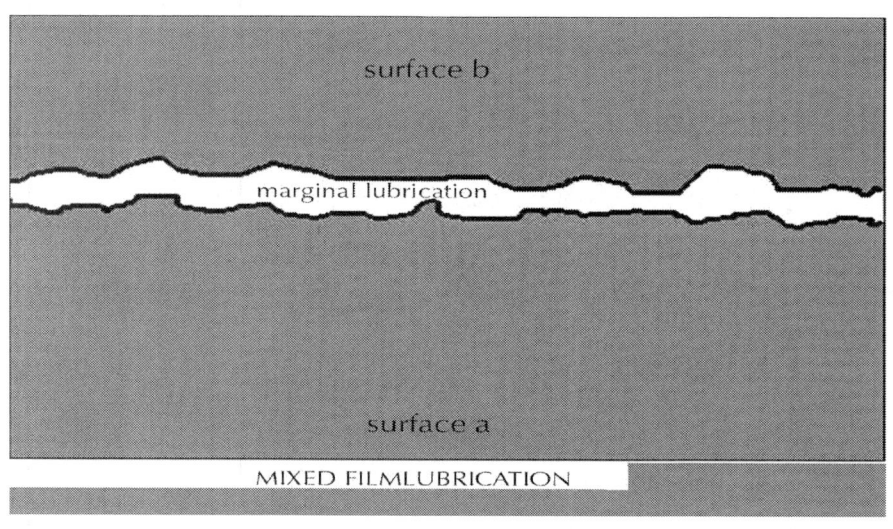

Scheme 1: Mixed-film lubrication.

Hydrodynamic Lubrication

Basically, lubrication is governed by one of two principles: hydrodynamic lubrication and boundary lubrication. In the former, a continuous full-fluid film separates the sliding surfaces. In the latter, the oil film is not sufficient to prevent metal-to-metal contact. Hydrodynamic lubrication is the more common, and it is applicable to nearly all types of continuous sliding action where extreme pressures are not involved. Whether the sliding occurs on flat surfaces, as it does in most thrust bearings, or whether the surfaces are cylindrical, as in the case of journal (plain or sleeve) bearings, the principle is essentially the same.

It would be reasonable to suppose that, when one part slides on another, the protective oil film between them would be scraped away. Except under some conditions of reciprocating motion, this is not necessarily true at all. With the proper design, in fact, this very sliding motion constitutes the means of creating and maintaining that film.

In zone 3 is the zone of hydrodynamic or fluid film lubrication where there is no wear because there is no contact between the surfaces. Hydrodynamic Lubrication is often referred to as stable lubrication. There are four essential elements in hydrodynamic lubrication, a liquid, relative motion, the viscous properties of the liquid, and the geometry of the surfaces between which the convergent wedge of fluid is produced. Only friction present in a hydrodynamic lubrication system is the friction of the lubricant itself, it would make sense to have a less viscous fluid in order to minimize friction: the less viscous a liquid the lower the friction. Too low of a viscosity jeopardizes our system though. We have to be very careful that the distance between the two surfaces is greater than the largest surface defect. The distance between the two surfaces decreases with higher loads on the bearing, less viscous fluids, and lower speeds. The surface geometry is also very important. The surfaces have to be such that a converging wedge of fluid can develop between the surfaces, allowing the hydrodynamic pressure of the lubricant to support the load of the shaft or moving surface. Hydrodynamic lubrication is an excellent method of lubrication since it is possible to achieve coefficients of friction as low as 0.001, and there is no wear between the moving parts. Special attention must be paid to the heating of the lubricant by the frictional force since viscosity is temperature dependent. One method of accomplishing this is to cycle the lubricant through a cooling reservoir in order to maintain the desired viscosity of the fluid. Another way of handling the heat dissipation is to use commercially available additives to decrease the viscosity's temperature dependence which are known as viscosity index improvers.

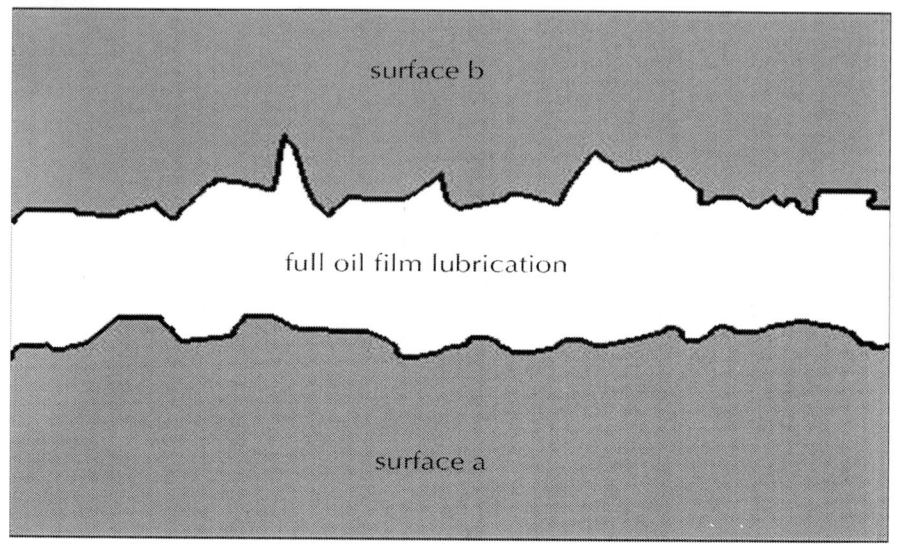

surface b

full oil film lubrication

surface a

HYDRODYNAMIC LUBRICATION

Scheme 2: Hydrodynamic lubrication.

The formation of fluid film is influenced by the following factors:

- The contact surfaces must meet at a slight angle to allow formation of the lubricant wedge.
- The fluid viscosity must be high to maintain adequate film thickness to separate the contacting surfaces at operating speeds.
- The fluid must be adhering to the contact surfaces for conveyance into the pressure area to support the load.
- The fluid must be distributing itself completely within the bearing clearance area.
- The operating speed must be sufficient to allow formation and maintenance of the fluid film.
- The contact surfaces of bearings and journals must be smooth and free from sharp surfaces that will disrupt the fluid film.

Boundary Lubrication

The oil film has become so thin in Zone 1 that there is no hydrodynamic contribution and only boundary lubrication which is defined by Campbell in 1969 as the lubrication by a liquid under conditions where the solid surfaces are so close together that appreciable contact between opposing asperities is possible. The friction and wear in boundary lubrication are determined predominantly by interaction between the solids and between the solids and the liquid. The bulk flow properties of the liquid play little or no part in the friction and wear behavior.

As mentioned, boundary lubrication is effective when a complete fluid film does not develop between potentially rubbing surfaces, the film thickness may be reduced to permit momentary dry contact between wear surface high points or asperities. Boundary lubrication occurs whenever any of the essential factors that influence formation of a full fluid film are missing. The most common example of boundary lubrication includes bearings, which normally operate with fluid film lubrication but experience boundary lubricating conditions during routine starting and stopping of equipment. Other examples include gear tooth contacts and reciprocating equipment.

A brief explanation of what needs to be added to basic mineral oil in order to create an effective boundary lubricant. Generally, the best additives are active organic compounds with long chain molecules and active end groups. These compounds bind tightly and intricately with each other, forming a film that builds up on the surface of the metal itself. This results in a thin film that is very difficult to penetrate. When two surfaces, each covered with a boundary layer, come in contact with each other they tend to slide along their outermost surfaces, with the actual faces of the surfaces rarely making contact with each other. Liquids are rarely good boundary lubricants. The best boundary lubricants are solids with long chains of high inter-chain attraction, low shear resistance so as to slip easily, and a high temperature tolerance. The boundary lubricant should also, obviously, be able to maintain a strong attachment to the surfaces under high temperatures and load pressures.

The most common boundary lubricants are probably greases. Greases are so widely used because they have the most desirable properties of a boundary lubricant. They not only shear easily, they flow. They also dissipate heat easily; form a protective barrier for the surfaces, preventing dust, dirt, and corrosive agents from harming the surfaces

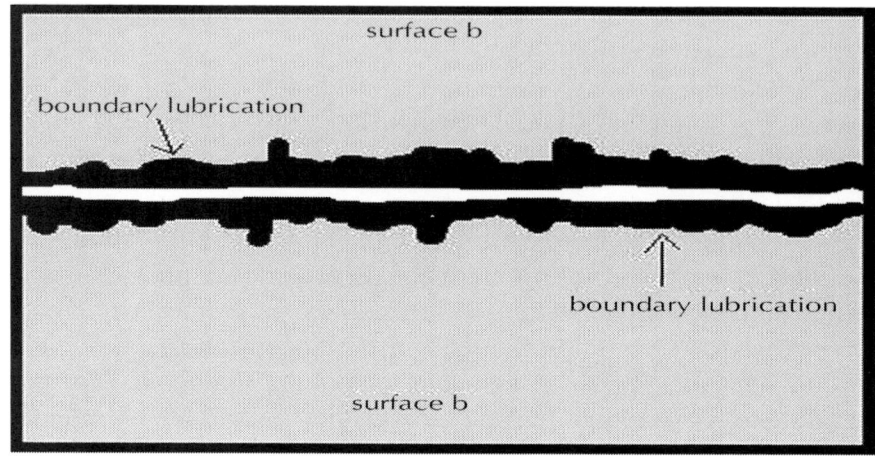

BOUNDARY LUBRICATION

Scheme 3: Boundary lubrication.

Base Stock

Petroleum is one of the naturally occurring hydrocarbons that frequently include natural gas, natural bitumen, and natural wax. The name "petroleum" is derived from the Latin petra (rock) and oleum(oil). According to the most generally accepted theory today, petroleum was formed by the decomposition of organic refuse, aided by high temperatures and pressures, over a vast period of geological time.

Although petroleum occurs, as its name indicates, among rocks in the earth, it sometimes seeps to the surface through fissures or is exposed by erosion. The existence of petroleum was known to primitive man, since surface seepage, often sticky and thick, was obvious to anyone passing by prehistoric animals were sometimes mired in it, but few human bones have been recovered from these tar pits. Early man evidently knew enough about the danger of surface seepage to avoid it.

The petroleum remaining from the distillation is thick like pitch; if the distillation has been pushed far, the residuum will flow only languidly in the retort, and in cold weather it becomes a soft solid, resembling much the maltha or mineral pitch Fig. 4 shows that the distillation of crude oil.

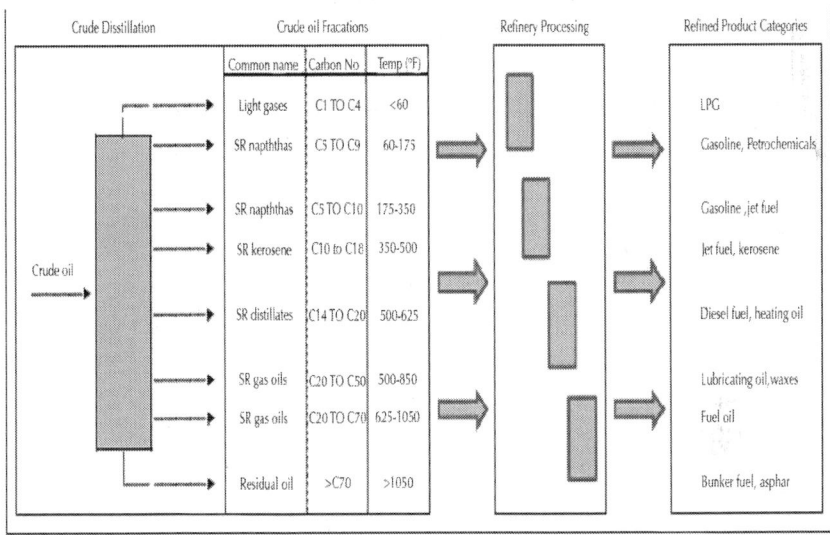

Crude Disstillation		Crude oil Fracations			Refinery Processing	Refined Product Categories
		Common name	Carbon No	Temp (°F)		
		Light gases	C1 TO C4	<60		LPG
		SR napththas	C5 TO C9	60-175		Gasoline, Petrochemicals
		SR napththas	C5 TO C10	175-350		Gasoline ,jet fuel
		SR kerosene	C10 to C18	350-500		Jet fuel, kerosene
Crude oil		SR distillates	C14 TO C20	500-625		Diesel fuel, heating oil
		SR gas oils	C20 TO C50	500-850		Lubricating oil,waxes
		SR gas oils	C20 TO C70	625-1050		Fuel oil
		Residual oil	>C70	>1050		Bunker fuel, asphar

Figure 4: Schematic view of crude oil distillation.

Base stocks are refined from crude oil to obtain products with the best lubricating properties. Base stocks generally make up 80-95% of a typical engine oil and 5% additives [1]. Base stock is used to describe plain mineral oil. The physical properties of an oil depend on its base stock. In most cases it is chemically inert there are three sources of base stock: biological, mineral and synthetic. The oils manufactured from these sources exhibit different properties and they are suitable for different applications. For example:

- Biological oils are suitable in applications where the risk of contamination must be reduced to a minimum, for example, in the food or pharmaceutical industry. They are usually applied to lubricate kilns, bakery ovens, etc. There can be two sources of this type of oil: vegetable and animal. Examples of vegetable oils are: castor, palm and rape-seed oils while the examples of animal oils are: sperm, fish and wool oils from sheep (lanolin).

- Mineral oils are the most commonly used lubricants throughout industry. They are petroleum based and are used in applications where temperature requirements are moderate. Typical applications of mineral oils are to gears, bearings, engines, turbines, etc.

- Synthetic oils are artificially developed substitutes for mineral oils. They are specifically developed to provide lubricants with superior properties to mineral oils. For example, temperature resistant synthetic oils are used in high performance machinery operating at high temperatures. Synthetic oils for very low temperature applications are also available [2].

Lubricants

All liquids will provide lubrication of a sort, but some do it a great deal better than others. The difference between one lubricating material and another is often the difference between successful operation of a machine and failure. For almost every situation, petroleum products have been found to excel as lubricants. Petroleum lubricants stand high in metal-wetting ability, and they possess the body, or viscosity characteristics, that a substantial film requires, these oils have many additional properties that are essential to modern lubrication, such as good water resistance, inherent rust-preventive characteristics, natural adhesiveness, relatively good thermal stability, and the ability to transfer frictional heat away from lubricated parts. What is more, nearly all of these properties can be modified during manufacture to produce a suitable lubricant for each of a wide variety of applications. Oils have been developed hand-in- hand with the modern machinery that they lubricate; indeed, the efficiency, if not the existence, of many of today's industries and transportation facilities is dependent upon petroleum lubricants as well as petroleum fuels.

The basic petroleum lubricant is lubricating oil, which is often referred to simply as "oil." This complex mixture of hydrocarbon molecules represents one of the important classifications of products derived from the refining of crude petroleum oils, and is readily available in a great variety of types and grades.

Any description of lubricating oils would be incomplete without consideration of oils for vehicle engines. These oils are used in greater quantity than all other lubricants combined, and are of interest to more people than any other lubricants. Engine oils are generally recommended by automotive builder according to the Society American of Automotive Engineers (SAE) viscosity classification.

Engine oil lubricants make up nearly one half of the lubricant market and therefore attract a lot of interest. The principal function of the engine oil lubricant is to extend the life of moving parts operating under many different conditions of speed, temperature, and pressure. At low temperatures the lubricant is expected to flow sufficiently in order that moving parts are not starved of oil. At higher temperatures they are expected to keep the moving parts apart to minimize wear. The lubricants reduce friction and removing heat from moving parts.

General Classification of the Lubricating Oils

The term lubricating oil is generally used to include all those classes of lubricating materials that are applied as fluids [3]. Lubricating oils are made from the more viscous portion of the crude oil which remains after removal by distillation of the gas oil and lighter fraction [4-8]. Although crude oils from various parts of the world differ widely in properties and appearance, there is relatively little difference in their elemental analysis. Thus, crude oil samples will generally show carbon content ranging from 83% to 87 %, and hydrogen content from 11% to 14%. The remainder is composed of elements such as oxygen, nitrogen, and sulfur, and various metallic compounds. An elemental analysis, therefore, gives little indication of the extreme range of physical and chemical properties that actually exists, or of the nature of the lubricating base stocks that can be produced from a particular crude oil.

An idea of the complexity of the lubricating oil-refining problem can obtained from a consideration of the variations that can exist in a single hydrocarbon molecule with a specific number of carbon atoms. For example, the paraffinic molecule containing 25 carbon atoms has 52 hydrogen atoms. This compound can have about 37,000,000 different molecular arrangements [3]. The hydrocarbons of the crude oils are:

Paraffinic Components

The paraffinic components, show in Fig. 5 (a, b), which determine the pour point, contain not only linear but also branched paraffins. The straight chain paraffins of high molecular weights raise the pour point of oils (waxy compounds) and should be removed by dewaxing processes.

The branched paraffins are chemically interesting hydrocarbons and they are found in large quantities in lubricating oil fractions from paraffinic crudes. Oil rich in paraffinic hydrocarbons have relatively low density and viscosity for their molecular weight and boiling range. Also, they have good viscosity/ temperature characteristics. In general, paraffinic components are reasonably resistant to oxidation and have particularly good response to oxidation inhibitors [9, 10].

Naphthenic Components

They have rather higher density and viscosity for their molecular weight compared to the paraffinic components. An advantage which naphthenic components have over the paraffinic ones is that they tend to have low pour point and so do not contribute to wax. However, one disadvantage is that they have inferior viscosity/ temperature characteristics. Single ring alicyclics with long paraffinic side chains, however, share many properties with branched paraffins and can in fact be highly desirable components for lubricant base oils. Naphthenic components, Fig. 5 (c), tend to have better solvency power for additives than paraffinic components but their stability to oxidative processes is inferior [9,10].

Aromatic Components

They have densities and viscosities which are still higher viscosity/ temperature characteristics are in general poor but pour point is low, although they have the best solvency power for additives, their stability to oxidation is poor. As for alicyclics, single ring aromatics with long paraffinic side chain may be very desirable base oil components, Fig. 5 (d). The classifying of hydrocarbon as paraffinic, naphthenic and aromatic groups which are generally used for characterizing the base oil should not be taken as absolute but as an expression of the predominating chemical tendencies of the base stocks [11].

Non Hydrocarbon Components

The non-hydrocarbons in lubricating oil are analogous in many ways to the hydrocarbons. Sulfur and nitrogen compounds are found almost entirely in ring structures such as sulfides, thiophene, and pyridine and pyrrol types. More complex molecules are also thought to exist in lubricating oil in which nitrogen and sulfur atoms are found in the same molecule. As in the case of hydrocarbons, these compounds will probably also have paraffinic side chains and possibly be condensed with naphthenic and aromatic ring structures [11]

Although these non-hydrocarbons may be present in only trace amounts, they often play a major role in controlling the properties of lubricating oils. In general they are chemically more active than the hydrocarbon, and hence they may markedly affect properties such as oxidation stability, thermal stability and deposit forming tendencies. In refining the general tendency is to reduce the non-hydrocarbons content to a minimum.

Naphthenic acid account for most of the oxygenated compounds found in petroleum. These are removed in the refining processes by neutralization and distillation. The naphthenates are retained in the residue from the distillation and can be removed by deasphalting process. Modern refining methods generally remove most of resins, asphaltenes, polycyclic aromatic, di aromatic and their analogous non hydrocarbons, so that the final lubricant consists chiefly of saturated and monocyclic aromatic fraction [12].

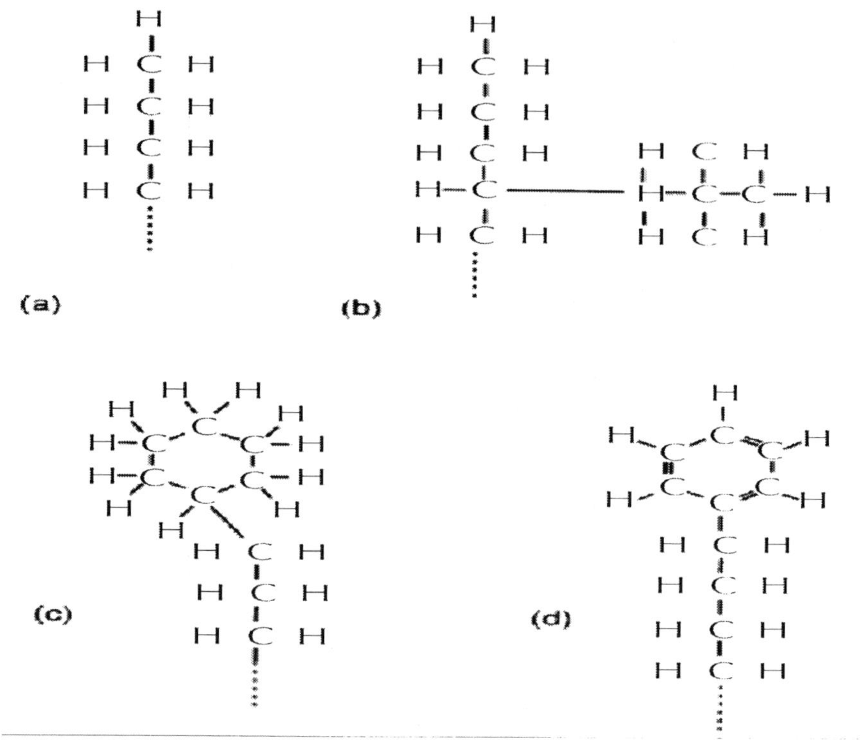

Figure 5: Chemical Structure of Lubricating oil.

Main properties of lubricating oils

The main properties which a lubricating oil must possess to full performance are:

Physical Properties of Lubricating Oil

- Viscosity: Viscosity is the measure of the internal friction within a liquid; the way the molecules interact to resist motion. It is a vital property of a lubricant because it influences the ability of the oil to form a lubricating film or to minimize friction [8]. Newton defined the absolute viscosity of a liquid as the ratio between the applied shear stress and the resulting shear rate.

- Viscosity index: The most frequently used method for comparing the variation of viscosity with temperature between different oils by calculation of dimensionless numbers, known as the viscosity index (VI). The kinematic viscosity of the sample is measured at two different temperatures (40°C, 100°C) and the viscosity compared with an empirical reference scale. VI is used as a convenient measure of the degree of aromatics removal during the base oil manufacturing process, but comparison of VI of different oil samples is only realistic if they are derived from the same distillate feedstock [8].

- Low temperature properties: When a sample of oil is cooled, its viscosity increases in a predictable manner until wax crystals start to form. The matrix of wax crystals becomes sufficiently dense with further cooling to cause an apparent solidification of the oil. Although the solidified oil does not pour under the influence of gravity, it can move if sufficient force is applied. Further decrease in temperature cause more wax to form, increasing the complexity of the wax/oil matrix. Many lubricating oils have to be capable of flow at low temperatures and a number of properties should be measured.

- cloud point: It is the temperature at which the first sign of wax formation can be detected. A sample of oil is warmed sufficiently to be fluid and clear. It is then cooled at a specified rate. The temperature at which haziness is first observed is recorded as the cloud point, the ASTM D 2500/IP 219 test. The oil sample must be free of water because it interferers with the test.

- pour point: It is the lowest temperature at which the sample of the sample of oil can make to flow by gravity alone. The oil is warmed and then cooled at a specified rate. The test jar is removed from the cooling bath at intervals to see if the sample is still mobile. The procedure is repeated until movement of the oil doesnot occur, ASTM D 97/IP 15. The pour point is the last temperature before the movement ceases, not the temperature at which solidification occurs. This is an important property of diesel fuels as well as lubricant base oils. High- Viscosity oils may cease to flow at low temperatures because their viscosity becomes too high rather than because of wax formation. In these cases, the pour point will be higher than the cloud point.

- High temperature properties: The high temperature properties of oil are governed by distillation or boiling range characteristics of the oil.
- volatility: It is important because it is an indication of the tendency of oil to be lost in service by vaporization.
- flash point: It is important for oil from a safety point of view because it is the lowest temperature at which auto-ignition of the vapour occur above the heated oil sample. Different methods are used, ASTM D 92, D93, and it is essential to know which equipment has been used when comparing results.
- Other physical properties: Various other physical properties may be measured, most of them relating to specialized lubricant applications. Some of the more important measurements are:
- density: Important, because oils may be formulated by weight, but measured by volume.
- demulsification: Ability of oil and water to separate.
- foam characteristics: Tendency to foam formation and stability of the foam that results.

pressure/viscosity characteristics

thermal conductivity: Important for heat transfer fluid.

- electrical properties: Resistively, dielectric constant.

surface properties: As surface tension, air separation.

Chemical Properties of Lubricating Oils

- Ease of starting rapidity of warming up.: The ease of starting depends chiefly on the cranking speed which is influenced by oil viscosity at the temperature of the crankcase. The major factor in the usage of a lubricant is its viscosity. It's not enough that the lubricants should have the proper viscosity but also they should maintain the little viscosity change within the temperature range during and after the appertain. So, viscosity controls not only frictional and thermal effect but also oil flow as a function of the load speed, temperature and design of the device lubricated. In other words, if the equipment will often have no make a cold start, it's also important that the viscosity at starting temperature is not so high that the machine cannot be started. The rapidity with which an engine can be put to work is dependent on the speed of circulation and supply of oil to vital components, all forms of wear and even the safety of the engine are influenced by rapidity of circulation of the lubricants.

- Low Carbon Forming Tendency.: This property is important for high compression ratio petro engine where carbon deposit will adversely affect combustion quality. The extent and also the composition of such formed deposits are causing noisy and rough burning which subjects the engine to high thermal and mechanical stresses resulting in lowering of performance and reduction of engine life. The typical symptoms will be knocking, preignition and surface ignition. These call higher octane fuels which are more expensive and do not eliminate the need for ultimate decarbonizing.

- Carbon residue test methods.: Provide with some indication about the relative coke forming tendency of the oil in some application and quality-controlled lubricants. So, the test can be helpful in selecting oils for certain industrial applications such as heat treating, lubrication of bearing subjected to high temperature and air compressors. It is claimed that the presence of viscous oil (bright stock) in the base oils plays an important role in the formation of carbon deposits.

- High Oxidation Stability.: One of the most important requirement of the lubricant is that its properties are not changed during use [5-10]. The lubricant is often subjected to several oxidizing conditions which are primarily due to the oxidative changes of the oil. While the temperature of the oil, engine parts presence of oxygen, nature by products of fuel composition contribute to the oxidative change the properties of the lubricant during use. Therefore, it's essential that the lubricating oil; when exposed to high temperature; doesn't contribute to the forming of deposits even after a long period of continuous engine running. So, the lubricant resistance to the oxidative depends mainly on the nature of the lubricant and the presence of anti-oxidant additives.

- Wear Reduction.: Wear occurs in lubricated systems by three mechanisms (abrasion, corrosion and metal-to-metal contact. i.e adhesion). The lubricant play an important role in combating each type of wear.

- Abrasive wear: It is caused by solid particles entering into the area between the lubricated surfaces and physically eroding these surfaces and may contaminate wear fragments. To cause wear, the solid particles must be larger than the oil-film thickness and harder than the lubricated surfaces. The flushing action of the lubricant, especially in forced feed or once through systems, severs to remove potentially harmful solid particles from the area of lubricated surfaces.

- Corrosive wear: Corrosive wear is generally caused by the products of oxidation of lubricants. The high sulfur content of the fuel helps the corrosive attack. In other words, corrosion is the principal cause of wear in the internal combustion engines because the products of combustion are highly acidic and contaminate the lubrication oil, lubricants function to minimize corrosive wear is in two ways: proper refinement plus the use of oxidation inhibitors which reduces lubricant deterioration and keeps the level of corrosive oxidation products low.

- Adhesive wear: This type of wear can significaly affect certain parts of the engine where metal-to-metal contact takes place. Adhesive wear takes place also if power was increased without corresponding modification is design, finishing and composition of the metal parts. Wear of this type also results from breakdown of lubricant film. It can also be the result of excessive surface roughness or interruption of the lubricant supply. A plentiful supply of the proper viscosity of oil is often the best way to avoid these conditions. The composition of the base oil and addition of certain chemical additives are also the important factors in protection of engine parts components against adhesive wear.

- Detergency and Dispersancy: With the exception of detergency and dispersancy in the combustion chamber, deposit in the oil are controlled by its detergent power. The source of the deposits found in engines are many and their volume depends mainly on the used, the quality of combustion, the temperature of lubricating oil and coolant, and on the gas sealing of the ring in the cylinder. It these deposits are not removed with the oil when it is drained, their accumulation in the engine would drastically shorten the engine life. The role of the detergent additives is to reduce the amount of deposits formed and their removal easy. The detergent property imparted to oils by additives seems to perform differently depending upon whether deposits result from high low temperature, low temperature deposits are mainly yielded from the fuel combustion and the detergency function is to keep them in suspension or solution in the lubricating oil. However, high temperature deposits are mainly related to the oxidized fraction of the oil. The role of detergency here is not only to maintain these products in suspension, but also to stop the development of those chain reaction which promote the formation of varnishes and lacquers. The physical and functional properties of the lube oil will depend on the properties of carbon atoms in the various ring structures and aliphatic side chain.

- Seal compatibility: Lubricants are often used in machines where they come into contact with rubber or plastic seal. The strength and degree of swell of these seals may be affected by interaction with the oil. Various tests have been devised to measure the effect of base oils different seals and under different test conditions [13]. The strength and degree of swell of these seals may be affected by interaction with the oil. Various tests measure the effects of base oils on different seals and under different test conditions.

Required Performance Characteristics for Lubricating Oils

Selection and application of lubricating oil are determined by the functions which are expected for performance. In one application, such as delicate instrument bearing, the reduction friction is paramount and in another, such as metal cutting, the temperature control may be most important. A lubricating oil performance or requirement for a modern high speed engine should fulfill the following five important functions:

- Reduction of the frictional resistance: The reduction of engine resistance to minimum is necessary to ensure maximum mechanical efficiency (running costs of a vehicle or engines are influenced by the lubricant viscosity)
- Protection of the engine against all types of wear: All users wants minimum maintenance costs, longer engine life and increased usefulness. Modern oil has allowed longer intervals between engines over hauls.

- Reduction of gas and oil leakages: The reduction of gas and oil leakages in an efficient and lasting manner is necessary to maintain engine performance and to prevent the combustions products from adulterating the oil.

- Contributing the thermal equilibrium of the engine: In modern engines, the oil functions and more as a heat exchange medium, dissipating the heat is not converted into work. This is often associated with the first function in this list where the viscous oil give greater frictional resistance and its slow internal circulation leads to a rapid temperature raise of some vital part of the engine to cool efficiency, the oil must be able to circulated quickly.

- Removal of all injurious impurities: The lubricant give the function of protecting the engine against corrosive and mechanical wear which caused by all injurious impurities. So, the removal of these impurities by lubricants is very important for engine. The function and the corresponding qualities required for engine lubricating oils are summarized in Table (1).

Table 1: Function and qualities required for engine oils

Main functions required	Qualities required
Reduce frictional resistance	• Viscosity not too high to provide good pumpability or to cause undue cracking resistance. • Minimum viscosity without risk of metal to metal contact under the varying condition of temperatures, speed and load. • Sufficiently high viscosity a high temperature; good lubrication property outside the hydrodynamic condition. • Anti-seizure properties, especially during the run-in period.

Protect against corrosion and wear	• Must protect metallic surface against corrosive action of fuel decomposition product (wear, SO_2, HBr, HCl, etec.)
	• Must resist degradation (resist oxidation and have a good thermal stability).
	• Must counteract action of fuel and lubricant decomposition product at high temperatures, especially on non-ferrous metals.
	• By intervention in the friction mechanism, must reduce the consequences of unavoidable metal-to-metal contact.
	• Must resist deposit formations which would affect lubrication (detergency or dispersancy action).
	• Must contribute to the elimination of dust and other pollutants (dispersancy action).
Assist sealing	• Must have sufficient viscosity at high temperature and low volatility.
	•Mustlimitwear
	• Must not contribute to formation of deposits and fight against such formation.
Contribute to cooling	• Must good and thermal stability and oxidation resistance.
	• Must have low volatility.
	• Viscosity must not be too high.
Facilitate the suspension andeliminate undesirable products	• Must be able to maintain in fine solid material whatever the temperature and physical and chemical condition.

Types of Lubricants

• Gaseous Lubricants: Gaseous lubricants belong to the simplest, lowest viscosity lubricants known and include air, nitrogen, oxygen, and helium. They are applied in aerodynamic and aerostatic bearings. Since the chemical properties and the aggregate state of most gases remain unchanged over a wide temperature range, gaseous lubricants offer several advantages over liquid lubricants. First, they can be applied at both very high and very low temperatures. Their chemical stability eliminates any risk of contamination of the bearing by the lubricant, important for the machinery used in many branches of industry, primarily in the food, pharmaceutical and electronic industries.

A useful property of gases is that their viscosities increase with temperature, wheras the opposite is true of liquids, resulting in load – carrying capacity of gas – lubricated bearings increasing with temperature. However, the relatively low viscosity of gases generally limits the load-carrying capacity of self-acting, aerodynamic bearings to 15-20kPa. It is possible to achieve better bearing performance with gaseous lubricants than with liquid lubricants due to the very low viscosity of the gases which results in smaller heat generation by internal friction. In some cases, such as in foil air bearings, sliding contact occurs during stops and starts [14], therefore solid lubricants such as PTFE are used to reduce friction.

- Liquid Lubricants: Mineral oils: As the hydrodynamic behavior of plain bearings of plain bearings is totally dependent on the viscosity characteristics of the lubricant, typical liquid bearing lubricants are straight mineral oil raffinates of various viscosity grades. The viscosity grade required is dependent upon bearing speed, oil temperature and load. Table (2) provides a general guideline to selecting the correct ISO viscosity grade. The ISO grade number indicated is the preferred grade for the speed and temperature range. ISO 68- and 100- Grade oils are commonly used in indoor, heated applications, with32- grade oils being used for high-speed, 10.000 rpm, units and some outdoor low temperature applications. The higher the bearing speed, the lower the oil viscosity required and also that the higher the unit operating temperature, the higher the oil viscosity required. If vibration or minor shock loading is possible, a higher grade of oil than the one indicated in table (2) should be considered.

Table 2: Plain Bearing ISO viscosity grade selection

Bearing Speed (rpm)	Bearing / Oil Temperature (ºC)			
	0-50	60	75	90
300-1,500	-	68	100-150	-
1,800	32	32 - 46	68 -100	100
3,600	32	32	46 - 68	68 -100
10,000	32	32	32	32-46

Other methods for determining the viscosity grade required in an application are to apply minimum and optimum viscosity criteria to a viscosity – temperature plot. A third and more complex method is to calculate the oil viscosity needed to obtain a satisfactory oil film thickness.

The lubrication of bearings for machine tools usually requires mineral oils of ISO VG 46 or 68. For fast – running grinding spindles with plain bearings, mineral oils of ISO VG 5 or 7 are required, dependent on bearing clearance and speed. Bearings operating under high loads need lubricants of ISO VG 68 or 100. The service life of the bearing can be increased if the viscosity of the selected liquid lubricant at operating temperature exceeds the calculated optimum viscosity.

On the other hand, increased viscosity also increases operating temperature. In practice, therefore, the extent to which lubrication can be improved in this way is often limited. The chemical compositions of these oils differ from typical base oils in that they contain somewhat more aromatic hydrocarbons and heterocyclic compounds, which act as natural oxidation inhibitors. An increased viscosity for oils derived from the same crude oil does not significantly change their chemical composition; the difference generally lies with the increasing chain length of the paraffinic hydrocarbons, mostly isoparaffins, and in the aliphatic substituents of naphthenic and aromatic rings, together with a slight increase in the number of naphthenic and aromatic rings. More highly refined mineral oils and oxidation inhibitors are used for applications where higher temperatures or longer service periods require better ageing stabilites.

Synthetic lubricants: in practice, every synthetic oil of adequate viscosity and good viscosity-temperature behavior can be used as a bearing lubricant, e.g. polyglycols are very good bearing lubricants for mills and calenders in the rubber, plastics, textile and paper industries. However, in most cases the synthetic oils specifically developed for lubricating particular equipment are also used to lubricate its bearings. Although synthetic oils do not form a lubricant film under pressure as well as mineral oils and may not be effective bearing lubricants despite their higher temperature viscosity.

Biodegradable products: Biodegradable products of vegetable or animal origin are also considered for liquid lubrication, e.g. the effects of sunflower oil added to base oil on the performance of journal bearings. The use of vegetable oils as lubricants is likely to increase due to environmental and government requirements and is becoming increasingly important.

- Solid Lubricants: General description: bearings used under vacuum, at very high temperatures or under very high radiation cannot be lubricated by liquid lubricants or greases. For these and many other cases, solid lubricants are used, deemed to be any solid material used to reduce friction and wear between two moving surfaces. In general, the solid material is interposed as a film between sliding and /or rolling surfaces. Simply stated, an adequate solid material is required for the special lubrication requirements of extreme operating conditions, such as very high or very low temperatures over a wide range, e.g. -200 to 850°C, and corrosive atmospheres. Such materials normally have a layered crystalline structure which ensures low shear strength, thereby minimizing friction. The shear strength between the crystalline layers is weak and sets up a low and sets up a low friction mechanism by slippage of the crystalline layers under low shearing forces. Examples of layer-lattice solids are molybdenum disulphide, graphite, boron nitride, cadmium iodide and borax. Solid lubricants are used mainly in the form of powders or as bonded solid films.

A good solid film lubricant has strong adhesion to the bearing substrate material, full surface coverage and good malleability. It should also be chemically stable and prevent corrosion, taking into account operational and environmental conditions. Many solid film lubricants have poor wear resistance, since any breaks in the film are not self-healing, in contrast to the surface coating formed by a liquid lubricant. Advanced solid film lubricants perform reliably in many specific applications and much experience has been gained to better understand their limitations. The most commonly used disulphide, graphite, polytetrafluroethylene propylene.

Another group of materials, the self-lubricated materials, are related to solid lubricants and are particularly important for bearings. Their self-lubricating characteristics eliminate the need of grease or other lubrication and Gove improved performance under high temperature conditions. Graph alloy (Graphite/metal) alloys make use of special properties of graphite, the structure of which can be compared to a deck of cards with individual layers able to easily slide off. This phenomenon gives the material a self-lubricating ability matched by few other materials and allows for the elimination of grease or oil that would evaporate, congeal or solidify, causing premature failure. The graphite matrix can be filled with a variety of embedded lubricants to enhance chemical, mechanical and tribological properties to give a constant, low friction coefficient rather than just a surface layer, helping to protect against catastrophic failure. Lubrication is maintained during linear motion where lubricant is not frawn out and dust is not pulled in.

A recent development in solid bearing lubricants is micro – porous polymeric lubricants, MPL, where a polymer containing a continous microporous network has oil contained within the pores, which may include appropriate additives [14]. The oil content in the polymer can be more than 50% by weight and the microporous polymer acts as a sponage, releasing and absorbing oil when necessary.

Lubricant Impurities and Contaminants

- Water Content: Water content (ASTM D95, D1744, D1533, and D96) is the amount of water present in the lubricant. It can be expressed as parts per million, percent by volume or percent by weight. It can be measured by centrifuging, distillation and voltammetry. The most popular, although least accurate, method of water content assessment is the centrifuge test. In this method a 50% mixture of oil and solvent is centrifuged at a specified speed until the volumes of water and sediment observed are stable. Apart from water, solids and other soluble are also separated and the results obtained do not correlate well with those obtained by the other two methods. The distillation method is a little more accurate and involves distillation of oil mixed with xylene. Any water present in the sample condenses in a graduated receiver. Voltammetry method is the most accurate. It employs electrometric titration, giving the water concentration in parts per million.

Corrosion and oxidation behavior of lubricants is critically related to water content. An oil mixed with water gives an emulsion. An emulsion has a much lower load carrying capacity than pure oil and lubricant failure followed by damage to the operating surfaces can result. In general, in applications such as turbine oil systems, the limit on water content is 0.2% and for hydraulic systems 0.1%. In dielectric systems excessive water content has a significant effect on dielectric breakdown. Usually the water content in such systems should be kept below 35 [ppm].

- Sulphur Content: Sulphur content (ASTM D1266, D129, and D1662) is the amount of sulphur present in an oil. It can have some beneficial, as well as some detrimental, effects on operating machinery. Sulphur is a very good boundary agent, which can effectively operate under extreme conditions of pressure and temperature. On the other hand, it is very corrosive. A commonly used technique for the determination of sulphur content is the bomb oxidation technique. It involves the ignition and combustion of a small oil sample under pressurised oxygen. The sulphur from the products of combustion is extracted and weighed.

- Ash Content: There is some quantity of noncombustible material present in a lubricant which can be determined by measuring the amount of ash remaining after combustion of the oil (ASTM D482, D874). The contaminants may be wear products, solid decomposition products from a fuel or lubricant, atmospheric dust entering through a filter, etc. Some of these contaminants are removed by an oil filter but some settle into the oil. To determine the amount of contaminant, the oil sample is burned in a specially designed vessel. The residue that remains is then ashed in a high temperature muffle furnace and the result displayed as a percentage of the original sample. The ash content is used as a means of monitoring oils for undesirable impurities and sometimes additives. In used oils it can also indicate contaminants such as dirt, wear products, etc.

- Chlorine Content: The amount of chlorine in a lubricant should be at an optimum level. Excess chlorine causes corrosion whereas an insufficient amount of chlorine may cause wear and frictional losses to increase. Chlorine content (ASTM D808, D1317) can be determined either by a bomb test which provides the gravimetric evaluation or by a volumetric test which gives chlorine content, after reacting with sodium metal to produce sodium chloride, then titrating with silver nitride [14].

CONCLUSIONS

1. The technology of lubrication has been used from the ancient times, from the pyramid building where massive rock slabs are moved, up to present modern times.

2. The main purpose of lubrication is to reduce friction and wear in bearings or sliding components to prevent premature failure.

3. Adequate lubrication also helps to prevent foreign material from entering the bearings and guards against corrosion and rusting. Satisfactory bearing performance can be achieved by adopting the lubricating method that is most suitable for the particular application and operating conditions

4. A lubricant prevents the direct contact of rubbing surfaces and thus reduces wear. It keeps the surface of metals clean. Lubricants can also act as coolants by removing heat effects and also prevent rusting and deposition of solids on close fitting parts.

5. Lubricant is consisting of either oil or grease. Most grease is from animal fats or vegetable lard.

6. Lubricating oils are made from the more viscous portion of the crude oil which remains after removal by distillation of the gas oil and lighter fraction

7. There are three major types of lubricants: Gaseous lubricants e.g. air, helium, Liquid lubricants e.g. oils, water and Solid lubricants

e.g. graphite, grease, teflon, molybdenum disulphide etc. Liquid lubricant is the most commonly used lubricant because of its wide range of possible applications while gaseous and solid lubricants are recommended in special applications.

8. Lubricants do not persist working without additives.

9. Additives are chemical compounds added to lubricating oils to impart specific properties to the finished oils. Some additives impart new and useful properties to the lubricant; some enhance properties already present, while some act to reduce the rate at which undesirable changes take place in product during its service.

REFERENCES

1. Rudnick Leslie R., Ewa A. Bardasz, and Gordon D. Lamb; "Lubricant Additives: Chemistry and Applications", Marcel Dekker, pages 387-427, (2003).

2. Stachowiak Gwidon W.,and Andrew W. Batchelor; Engineering Tribology", third edition, Amsterdam: Elsevier, pages 2,12,-22,52,62-67,77, (2005).

3. Geore J.W.; "Lubrication Fundamentals", (1980).

4. Dowson D.; History of Tribology, 2nd Edition, Professional Engineering Publishing, London (1998).

5. Pirro D.M. and Wessol A.A.; "Lubrication fundamentals"; Marcel Dekker, Inc. New York and Basel 3, 37 (2001).

6. Spikes H.; Tribology International 34, 789 (2001).

7. Stachowiak G. W. and Batchelor A.w.; Engineering Tribology, 2nd Edition, Butterworth-Heinemann, Boston, (2001).

8. Pawlak Z., "Tribochemistry of Lubricating Oils"; Elsevier, UK, 45, 17 (2003).

9. Avilino S.Jr., "Lubricant Base Oil and Wax Processing", Morcel Dekker,Inc.,New York, Chapter (2),pp.17-36,(1994).

10. Mortier R.M. and Orszulik S.T.; "Chemistry and Technology of Lubricant", Blackie Academic and Professional Publications, Chapter (1), pp.2-12,(1993).

11. O'Connar J.J., Boyd J. and Auallane E.A.; "Standard Hand Book of Lubrication Engineering", McGrow Hill, New York, 14-2 (1968).

12. Allyson M., Keith D., Vincent R. and Thibon A.; Tribology International 34, 389-395 (2001).

13. Anon, Machinery and Production, 19 July 24 (1996).

14. Roy M. M., Malcolm F. F., and Stefan T. Orszulik; Chemistry and Technology of Lubricants, 3rd Edition, 12-13, (2010).

Chapter 3

New Scuffing Test Methods for the Determination of the Scuffing Resistance of Coated Gears

Remigiusz Michalczewski[1], Marek Kalbarczyk[1], Michal Michalak[1], Witold Piekoszewski[1], Marian Szczerek[1], Waldemar Tuszynski[1] and Jan Wulczynski[1]

[1]Institute for Sustainable Technologies - National Research Institute (ITeE-PIB), Tribology Department, Radom, Poland

INTRODUCTION

Scuffing of Gear Teeth

In modern machines the problems of the prevention of scuffing of the gear teeth is still very important. One of the reasons is that for many years the technique development is related to increasing the loading of the friction surfaces accompanied by decreasing their size

[1]. In the case of gears, the risk of scuffing occurrence rises because of potential design and assembly mistakes, unexpected overloads, as well as extremely different speeds of the rotation of gears, because both very high speeds and very low speeds may cause scuffing [2]. The occurrence of one of the mentioned factors may lead to very serious gear failures.

Apart from the above mentioned factors, the problems of using proper lubricating oils, with high extreme-pressure (EP) properties cannot be neglected.

In gears, the surface destroyed by scuffing appears at the addendum and dedendum of the tooth. This results from the sliding speed of the meshing teeth that reaches the highest values at these places of the gear tooth. Failures of the gear teeth flanks due to scuffing are shown in Figure 1.

(a)   (b)

Figure 1: Photographs of failures of the gear teeth flanks due to scuffing: a) "non-symmetric" scuffing observed in gear service, resulting from the incorrect distribution of load along the tooth [3], b) scuffing on the flank of the test gear due to poor extreme-pressure (EP) properties of the tested gear oil during the gear scuffing experiments performed by the authors.

Another example of scuffing of gears concerns the rudder speed brake power drive unit of a space shuttle, observed during its inspection after grounding [4]. Figure 2 a) shows the pinion and ring gear of the power drive unit of the space shuttle. Figure 2 b) presents the pinion tooth with wear at the tip and scuffing on dedendum. It was postulated that early shutdown of one of three hydraulic motors driving the gearbox could cause scuffing - in a differential gearbox, early shutdown of one motor could cause the overloading with potential for scuffing.

(a)   (b)

Figure 2: Photographs of the components of the rudder speed brake power drive unit of a space shuttle: a) pinion and ring gear of the power drive unit, b) damaged pinion tooth [4]

From the above example, it is absolutely apparent that the prevention of scuffing is still an important challenge, even in the high-tech sector.

Scuffing — How is it Brought About?

To better understand scuffing, Figure 3 presents the interpretative models of the phenomena in different phases of this process, caused by the continuously increasing load. The models concern the contact between two balls of the four-ball tribosystem (the rotating upper ball with one of the three stationary lower balls) during the testing of the automotive gear oils of API GL-4 and GL-5 performance levels. Such oils contain chemically active extreme-pressure (EP) lubricating additives to prevent scuffing. API GL-4 oils are used to lubricate synchronised manual transmissions of European cars and contain up to 4% of EP additives. API GL-5 oils containing up to 6.5% of EP additives are employed to lubricate automotive gears especially susceptible to scuffing, i.e. hypoid gears, in axles operating under various combinations of high-speed/shock-load and low-speed/high-torque conditions.

It should be emphasised here that a four-ball tribosystem is very often used for tribological testing of the performance of automotive gear oils.

The lower graph in Figure 3 presents the friction torque curve (M_t) obtained at continuously increasing load (P). The brackets over the graph indicate particular phases of the scuffing process. In these phases, the friction coefficient values (μ) were determined, and they are given in the red rectangles in the graph area. The thick red line below the graph denotes the time from the beginning of the run until the occurrence of the scuffing initiation reflected by a sharp rise in the friction torque.

The interpretative models of phenomena related to scuffing are presented over the graph in Figure 3. Because the models concern the contact zone between two balls of the four-ball tribosystem where the upper ball rotates and the lower ball is stationary, the direction of the movement was indicated in the upper part of the models by an arrow. If there is no arrow, the given model illustrates no movement of the balls, i.e. at the beginning of the run (before the motor of the tribotester starts).

For the phase "scuffing initiation," the upper model in Figure 3 illustrates the surface that did not exhibit very rough topography typical of scuffing (shown in the surface topography image), while the lower one concerns the surface already destroyed by scuffing.

In the models, three characteristic zones in the wear scar surface layer were identified: a chemically modified zone through the action of the lubricating additives and the steel surface, a zone of plastic deformation, and a zone of elastic deformation. All of these zones are described in the legend above the models in Figure 3.

Phase: "Beginning of Run"

After immersing the test balls in the tested gear oil and applying the initial load close to 0, a phenomenon known as physical adsorption or physisorption appears. In this phase, adsorbed molecules constitute the boundary layer on the friction surface, which protects the surface asperities against direct contact. The model with the heading "Beginning of run" in Figure 3 illustrates this, reflecting the situation before the start of the relative movement of the test balls.

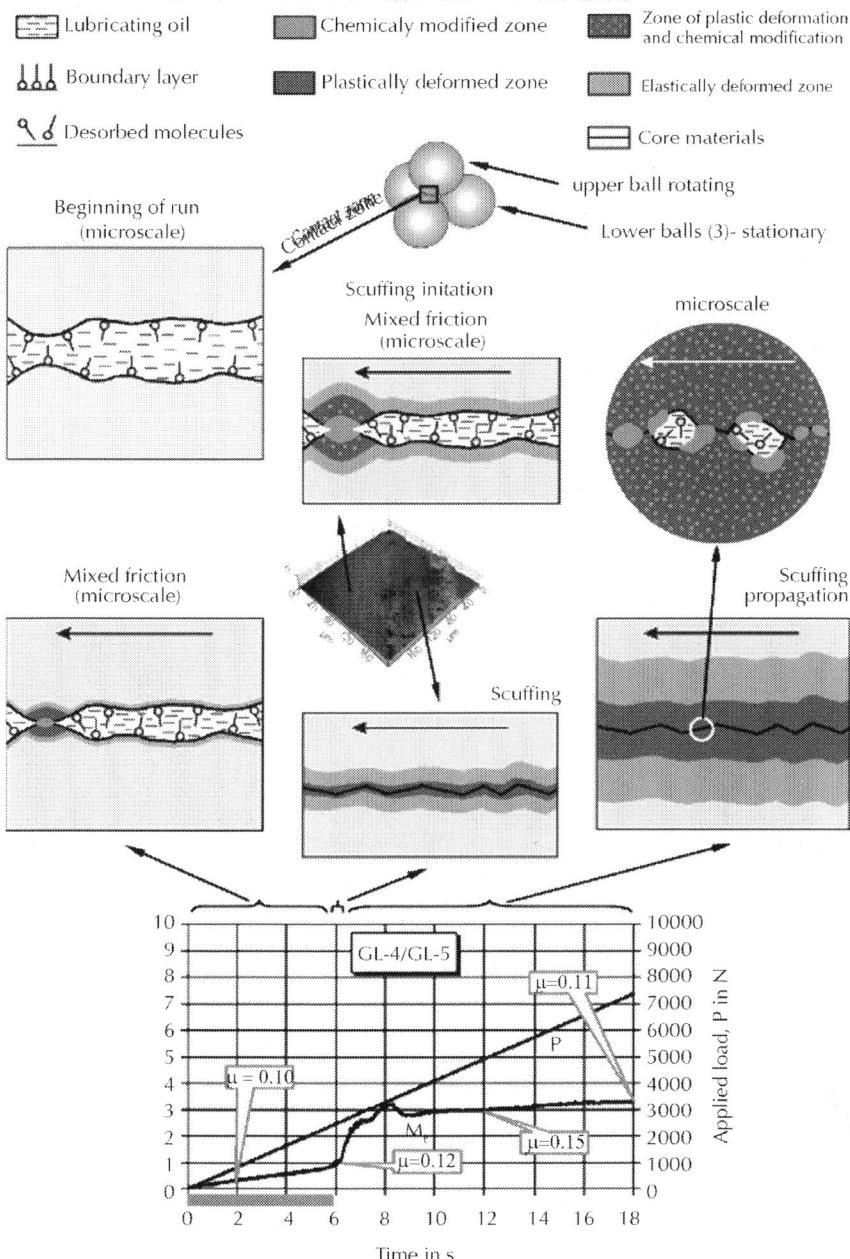

Figure 3: Models of scuffing in different phases for the automotive gear oils of the API GL-4 and GL-5 performance levels.

1.2.2. PHASE: "MIXED FRICTION"

In publications, the terms "mixed friction" and "mixed lubrication" are often used equivalently and concern the same phenomena. For the purpose of this chapter, one can assume that occurrences during the regime of the mixed lubrication result in the mixed friction with its specific friction coefficient.

The phase "Mixed friction" concerns the first stage of the run from the moment of the start of the relative movement between the test balls to the scuffing initiation reflected by a sharp rise in the friction torque. Its duration is denoted by the thick red line below the graph with the friction torque (M_t) and applied load (P) - Figure 3.

In this phase the mixed friction occurs. This can be stated on the basis of the fundamental criterion that is the friction coefficient value. The friction coefficients typical of particular types of friction were adopted from the work [5], where the four-ball tribosystem was also employed. From that work, it implies that the mixed friction occurs in the four-ball tribosystem when the friction coefficient is in the range between 0.07 and 0.1. Thus, the authors determined the friction coefficient at the 2nd second of the four-ball experiment, being 0.1, denote the mixed friction.

It is worth noting here that the idea of the occurrence of the mixed friction regime (instead of EHL, i.e. elastohydrodynamic lubrication) at the very start of the relative movement between the test balls (load is close to 0) is also supported in the mentioned work [5]. From that work it is apparent that "pure" EHL occurs in the four-ball tribosystem only under conditions of a low load and high speed.

At mixed friction, the micro-EHL films mainly carry the load and the mating surfaces are protected from direct contact by the boundary layer. But at some micro-zones, due to the failure of the micro-EHL film surface, asperities locally collide, which is illustrated in the model with the heading "Mixed friction" in Figure 3.

Due to collisions of surface asperities, the temperature in the micro-contact rises. At a higher temperature, physically adsorbed molecules may be attracted to the surface with greater forces, and chemical adsorption or chemisorption appears. The decomposition of the active compounds in the lubricating additives catalyses the transformation of some chemically adsorbed molecules into chemical compounds at higher temperatures.

The collision of the surface asperities and the local high pressure of the oil induced by the approaching asperities bring about elastic (reversible) and plastic (irreversible) deformations of the contacting surface. Due to the thermal (temperature rise) and mechanical activation (plastic deformation causing surface defects), the conditions exist for the initiation of the diffusion of "active" atoms from the lubricating compounds (e.g. sulphur atoms) into the surface layer.

The described phenomena lead to the formation of inorganic chemical compounds of iron with sulphur, phosphorus, and oxygen, coming from EP lubricating additives in the tested gear oil. Such additives (based on organic S-P compounds) form e.g. iron sulphide FeS [6]. FeS compounds, apart from hampering the creation of adhesive bonds with their shear strength being 1/5th that of steel and their hardness being 1/4th that of steel, facilitate shearing of the chemically modified surface asperities, and the shear plane is transferred to the thin FeS layer, which protects the surface from tearing out the material from deeper layers, reducing the wear intensity.

For the tested oil, containing EP lubricating additives, the surface asperities are covered by the protective layer of the above mentioned chemical compounds. This is illustrated in the respective model in Figure 3. Due to this, for the gear oils with EP lubricating additives, the scuffing initiation is delayed to appear at much higher loads than in the case of oils without lubricating additives (e.g. API GL-1 ones, not presented here).

Scuffing Phase: "Scuffing Initiation"

In this phase, scuffing initiates - the friction torque (M_t) sharply increases and measured friction coefficient values exceed the maximum value assumed for the mixed friction, i.e. 0.1 [5].

The scuffing initiation occurs at a load called the scuffing load, which is characteristic for each tested lubricating oil. At this load, the lubricating film collapses, the number of colliding surface asperities drastically increases, and the destruction changes its occurrence from the micro- to macro-scale and scuffing appears. Initially only part of the friction surface undergoes scuffing. It can be observed in the surface topography image of the border between the surface that did not exhibit very rough topography typical of scuffing (left side) and the surface destroyed by scuffing (right side) - Figure 3.

The described phenomena leading to scuffing are illustrated in the models with the heading "Scuffing initiation" in Figure 3. The upper model concerns the surface that did not exhibit very rough topography typical of scuffing, where still the mixed friction exists, while the lower one refers to the surface already destroyed by scuffing.

The upper model shows that the micro-scale phenomena in the zone intact by scuffing are similar to those described in the phase "Mixed friction" apart from the thickness of elastic and plastic deformations which increased due to rising load. Probably, in view of plastic deformation that causes surface defects, the reactive diffusion of "active" atoms from the EP lubricating additives (e.g. sulphur atoms) into the surface layer takes place and iron sulphides form, which is confirmed by other researchers, e.g. in the work [7]. The diffusively modified micro-zones inside the highest asperities are plastically deformed and are indicated in the respective model as orange spots - Figure 3.

By observing phenomena in the part of the friction surface that undergoes scuffing, one can indicate that the situation changes radically. The lower model illustrates that, in the first phase of scuffing, the lubricating film no longer exists, nor is there any boundary layer. This leads to a rapid intensification of the material destruction. Much plastic deformation appears, turning into the transfer, flowing and mingling of the material of the rubbing test balls. For the tested oils with EP lubricating additives, much of the surface layer starts to be chemically modified. This will be decisive for the scuffing propagation character.

Scuffing Phase: "Scuffing Propagation"

This phase refers to the scuffing process, after its initiation. It is reflected by a sharp increase in the friction torque (M_t), accompanied by a high intensity of the lower test balls wear - Figure 4 a, b). This situation is illustrated in the models with the heading "Scuffing propagation" in Figure 3.

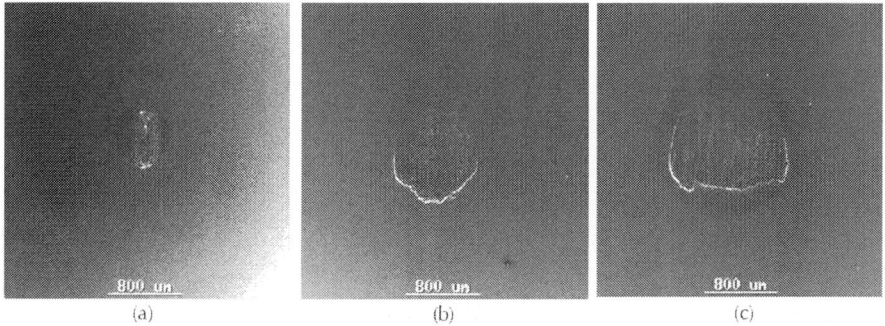

(a)   (b)   (c)

Figure 4: Development of the wear of the lower test balls due to scuffing: a) at scuffing initiation, b) at 12[th]seconds of the run (scuffing propagation), c) at the end of the run; images obtained at the same magnification.

For the tested gear oils, after the scuffing initiation due to rapid chemical reactions of their EP additives with the surface, a rise in the friction torque is mitigated to quickly stabilise at relatively low value -Figure 3. It is accompanied by continuously evolving wear of the lower test balls that is not intensive -Figure 4 b, c). A drop in the pressure in the contact zone due to wear, brings about the possibility of oil introduction into the contact zone and the regeneration of the boundary layer on much of the friction surface. Such an action is indicated by the friction coefficient within the range 0.11 to 0.15, typical of boundary friction. On the basis of the work [5], which also concerns four-ball experiments, it was assumed that the boundary friction occurs in the four-ball tribosystem when the friction coefficient is in the range between 0.09 and 0.15. The determined values of the friction coefficient being in the middle and upper limit typical of boundary friction denote that some part of the friction surface must have undergone scuffing; It can be assumed from [5] that "full scuffing" occurs when the friction coefficient exceeds 0.3. The specific state of the surface layer in this phase is called the "Secondary Boundary Layer" (SBL) in the work [8]. The round model in the micro-scale concerning the scuffing propagation (Figure 3) illustrates the places of oil appearance in the contact zone. Let us call them "the micro-pockets." One can presume that inside the oil micro-pockets the following phenomena take place: the intensive adsorption and desorption of the base oil and lubricating additives molecules on/from the steel surface, chemical reactions of the lubricating additives with the surface, and - in view of plastic deformation that causes surface defects - the diffusion of "active" atoms from the lubricating compounds (e.g. sulphur atoms) into the surface layer. In view of the transfer and mingling of the material of the rubbing test balls, the chemically modified zones appear across the entire zone of plastic deformations - orange spots. For the API GL-4 and GL-5 gear oils, the effective chemical modification of the surface mitigates the increase of the wear scar diameter - Figure 4 b, c) - in the phase of the SBL formation, accompanied by a mitigated rise in the friction torque and a decreasing friction coefficient (Figure 3).

Gear Tests of Scuffing

Nowadays, two manners of the improvement of the resistance to scuffing of gears are in use in the world. One is focused on the improvement of extreme-pressure (EP) properties of gear oils. The other one is related to the improvement of the properties of gear materials, e.g. by the deposition of thin hard coatings onto the tooth flank surface.

The verification of the quality of gear oils and new techniques of surface engineering of the tooth surface of gears requires that gear testing should be used. The most known is a unique complex of gear test methods developed in the Gear Research Center (FZG) at the Technical University of Munich. Approximately, 500 FZG gear test rigs are used around the world [9].

The most often used and popular gear tests for lubricating oils are performed using the FZG A/8.3/90 scuffing test method. Unfortunately, this method makes it impossible to differentiate between gear oils having very good extreme-pressure (EP) properties, from the point of view of the resistance to gear scuffing [10]. This is why various scientific centres have developed their own test methods [10-13].

Recognising the problem of the low resolution of A/8.3/90 scuffing test, the FZG has developed two new scuffing methods denoted as A10/16.6R/90 and S-A10/16.6R/90 (S - *shock*). The new test methods are described in detail in the literature, e.g. [14-19]. They are carried out under much severer conditions compared to the A/8.3/90 test. This is a result of the reduced face width of the small gear (pinion), doubling rotational speed, and reversing the direction of rotation. Additionally, according to the S-A10/16.6R/90 method, the test is started at once with a load at which the failure is expected, hence the name "scuffing shock test." Shock loading prevents the test gears from running-in and in turn increases their susceptibility to scuffing, which further increases the method resolution.

Nowadays, one of the research directions in numerous scientific centres in the world is an improvement in the scuffing resistance of toothed gears, achievable by the deposition of thin, hard, low-friction coatings onto the gear teeth, e.g. the a-C:H:W or MoS_2/Ti coatings [20-22]. For the last several years, intensive research work has also been performed on this subject in the Tribology Department of ITeE-PIB. Until now, the FZG A/8.3/90 gear scuffing test method has been used most often in various scientific centres, which, like in the case of testing gear oils, exhibits a resolution that is too low to differentiate between the coated gears from the point of view of their resistance to scuffing [23-25] - Figure 5. It should be explained here that a-C:H:W and a-C:H coatings are DLC (diamond-like carbon) coatings, and the a-C:H:W coating has an outermost DLC layer doped with W (tungsten).

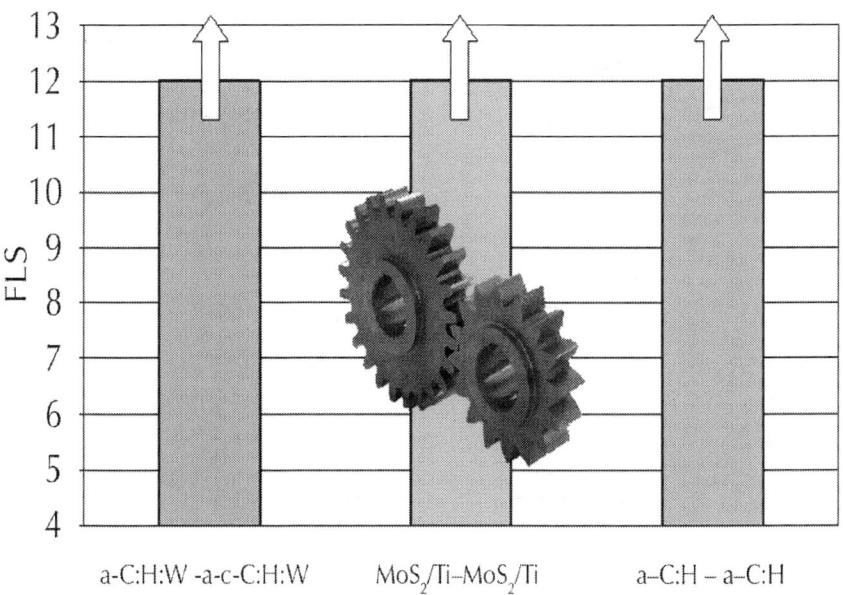

Figure 5: Failure load stages (FLS) obtained for the tested coatings (both gears coated) - FZG A/8.3/90 test method; data compiled from [23-25].

It is apparent from Figure 5 that the failure load stages (FLS), indicating the gear resistance to scuffing exceed the maximum number 12, so that the it is impossible to differentiate between the coated gears using the FZG A/8.3/90 test method.

To solve this problem, in the Tribology Department of ITeE-PIB, research was undertaken to apply the new FZG scuffing tests for coated gears to differentiate between their resistances to scuffing. Because the FZG test methods are dedicated exclusively to lubricating oils, their application for testing coated gears required introducing significant modifications - unique test methods have been developed, being the subject of this chapter. They are called the "Gear Scuffing EP Test for Coatings" and "Gear Scuffing Shock Test for Coatings."

NEW TEST METHODS

Idea of the Methods

The main difference between the test methods designed by the authors and the gear scuffing tests A10/16.6R/90 and S-A10/16.6R/90, developed by FZG, is a rise in the initial oil temperature to 120 °C, adoption of a failure criterion related to wear of the wheel (big gear), and resigning from the criterion of invalidation of the test results when wear of the wheel exceeds 20 mg.

The tests are performed on a pair of lubricated test gears with a coating (it can be applied on one or both the gears) at a constant rotational speed, and at the initial temperature of the lubricating oil identical for all the runs - until a failure load stage (FLS) is determined, i.e., such a load at which at least one of the failure criteria is met. In the Gear Scuffing EP Test for Coatings, based on the FZG S-A10/16.6R/90 test, the load is increased stepwise, from the lowest to the highest value. According to the Gear Scuffing Shock Test for Coatings, based on the FZG S-A10/16.6R/90 test the load is not increased in stages from the lowest value, but the expected failure load is applied to an unused gear flank (hence, the name "shock test"). In the shock test, each change of the load requires an unused gear flank; therefore, before subsequent runs, the test gears should be disassembled and reversed or replaced with new ones.

Although the authors have introduced some significant changes to the FZG gear scuffing tests, the core procedures of performing the tests are the same as in the FZG tests, and they can be found in the relevant publications, e.g. in [14].

To better explain the differences between the "old" FZG gear scuffing test A/8.3/90 and the new test methods designed by the authors, the test conditions according to each method and the failure criteria are specified in Table 1.

Table 1: Comparison of the FZG gear scuffing test and the methods designed by the authors

	Gear scuffing test FZG A/8.3/90	Gear Scuffing EP Test for Coatings based on FZG A10/16.6R/90	Gear Scuffing Shock Test for Coatings based on FZG S-A10/16.6R/90
Purpose of test	Testing lubricating oils	Testing coatings deposited on gears	Testing coatings deposited on gears
Test gear type	FZG A-type (pinion and wheel width 20 mm)	FZG A10-type (pinion width 10 mm, wheel width 20 mm)	FZG A10-type (pinion width 10 mm, wheel width 20 mm)
Test materials	20MnCr5	20MnCr5, but at least one gear coated	20MnCr5, but at least one gear coated
Motor rotational speed	1500 rpm	3000 rpm	3000 rpm
Circumferential speed	8.3 m/s	16.6 m/s	16.6 m/s
Direction of motor rotation	"Normal"	"Reversed" (R)	"Reversed" (R)
Run duration	15 min.	7 min. 30 s	7 min. 30 s
Maximum load stage	12	10	12
Maximum loading torque	535 N·m	373 N·m	535 N·m
Maximum Hertzian pressure	1.8 GPa	2.2 GPa	2.6 GPa

Loading type	Stepwise, from load stage 1	Stepwise, from load stage 1	Shock (i.e. starting with a load at which the failure is expected)
Initial lubricating oil temperature	90 °C	120 °C	120 °C
Temperature stabilisation during the run by cooling	No	No	No
Type of lubrication	Dip lubrication	Dip lubrication	Dip lubrication
Main failure criterion for FLS determination	$A_p \geq$ area of one pinion tooth ($\approx$200 mm^2)a	$A_p >$ area of one pinion tooth ($\approx$100 mm^2)	$A_p >$ area of one pinion tooth ($\approx$100 mm^2), or $W_w >$ 200 mgb
Additional criteria of failure assessment	None	Failures on the pinion teeth	Failures on the pinion teeth
Criterion of invalidation of the run	None	Significant decohesion of the coating	Significant decohesion of the coating

aA_p - total area of failures on the pinion

bW_w - wear (mass loss) of the wheel

After starting the run, the oil in the test chamber is heated by the heaters and friction. The oil temperature is allowed to rise freely. No cooling system is used in the tests.

Like in the FZG gear scuffing tests, if the failures are observed only within 1 mm from the tooth addendum, they are only scratches, or the failures are so small that the original criss-cross-grinding pattern (Figure 6) is still intact, they should be neglected when calculating the total area of the failures.

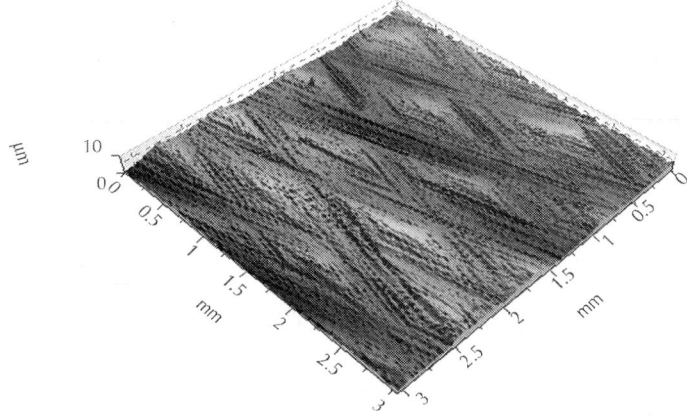

Figure 6: Original criss-cross-grinding pattern on the test gear teeth - stylus profilometry image.

The failure load stage (FLS) is the main measure of the resistance of the test gears to scuffing. According to the Gear Scuffing EP Test for Coatings, the FLS is such a load at which the main failure criterion specified in Table 1 has been met. According to the Gear Scuffing Shock Test for Coatings, the FLS is such a load at which at least one of the failure criteria has been met and, when at the load stage lower by 1, neither of the failure criteria has been met.

When there is significant decohesion of the coating due to poor adhesion to the surface, the run should be invalidated.

After run completion at a given load stage, the failures on the pinion teeth should be noted using the symbols from Table 2. These data are used for additional failure assessment, complementarily to FLS.

Table 2: Modes of wear of the test pinion (small gear)

Mode of wear	Symbol	Appearance
Polishing	W	
Scratches	R	
Scoring	B	
Scuffing	Z	

Polishing can be identified when the "mirror-like" surface on the tooth flank is observed with the disappearing criss-cross-grinding pattern shown in Figure 6.

Scratches appear as shorter or longer fine lines in the sliding direction of the tooth flanks.

Scoring marks run in the same direction as scratches. On the basis of CEC L-07-95 standard [26], it can be adopted that they occur singly or in zones as light, medium or deep grooves continuing towards the tip of the tooth and having a rougher appearance than the criss-cross-grinding pattern (Figure 6).

Scuffing marks occur as single, fine marks or strips, or areas covering a part or all of the flank width. According to CEC L-07-95 standard, they appear as dull areas with the roughness much greater than the original criss-cross-grinding pattern shown in Figure 6. In this case, the grinding pattern is no longer visible.

The difference between scuffing and scoring is that scuffing originates from the adhesive bond creation between the mating surfaces, which are then sheared, and scoring results from mechanical abrasion of the surface by the very hard wear particles under conditions of a very high load. Like scuffing, scoring is one of the most dangerous modes of gear wear.

When both test gears are uncoated, a respective standardised test method A10/16.6R/90 or S-A10/16.6R/90, developed by FZG should be used. However, to compare the results with the new test methods, it is necessary to start a run at the initial oil temperature of 120 °C rather than 90 °C.

Test Gears

A photograph of the FZG A10 scuffing test gears employed in the tests according to the developed methods is shown in Figure 7.

Figure 7: Photograph of the FZG A10 scuffing test gears.

The A10 test gears are made of 20MnCr5 steel. They are carburized, case hardened, tempered and Maag criss-cross ground. The surface hardness is HRC = 60 + 2 and the case hardness depth (CHD) is 0.6 to 0.9 mm (Eht). The effective face width of the pinion is 10 mm, and the wheel is 20 mm. The number of pinion teeth is 16, and wheel 24. The gears are identical to the ones used to perform tests according to the FZG A10/16.6R/90 and S-A10/16.6R/90 methods.

Test Equipment

For the complex testing of gears, a back-to-back gear test rig, denoted as T-12U, was designed in the Tribology Department of ITeE-PIB in Radom. Its photograph is presented in Figure 8 and kinematic schemes are presented in Figure 9.

The T-12U test rig is equipped with a control-measuring system, which consists of measuring transducers (thermocouple, speed transducer) and the controller (Figure 8).

Figure 8: Photograph of the T-12U gear test rig.

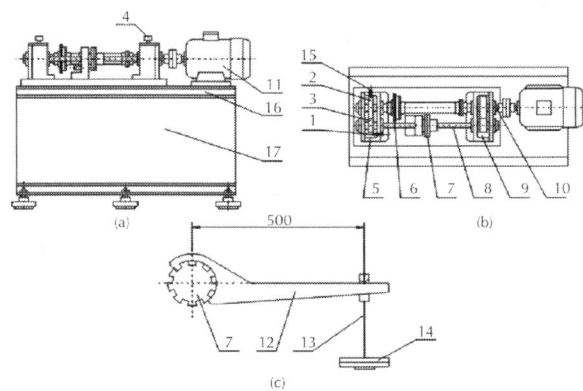

Figure 9: Kinematic schemes of the T-12U gear test rig: a) front view, b) top view, c) loading equipment; 1 - thermocouple, 2 - test wheel, 3 - test pinion, 4 - vent, 5 - test chamber, 6 - shafts torsion angle indicator, 7 - load clutch, 8 - front shaft, 9 - slave chamber, 10 - drive clutch, 11 - electric motor, 12 - loading lever, 13 - weight hanger, 14 - weights, 15 - heaters, 16 - frame, 17 - concrete base.

During runs, the following quantities are measured: rotational speed, lubricating oil temperature, motor current load, time, and the number of motor revolutions. The measured values are displayed on the controller.

The test rig is mounted on the concrete base equipped with vibration-dumping feet.

The T-12U gear test rig is a back-to-back rig (Figure 9) where the test gears (2) and (3), located in the test chamber (5), are connected by two shafts to the slave gears, located in the chamber (9). The front shaft (8) has two parts. Between them there is the load clutch (7). To apply the loading torque between the meshing gears, before the run, one part of the shaft (the left part of the front shaft (8) is fixed to the base with the lock-pin via the clutch and its support. A round-shaped loading lever (12) is placed on the right part of the clutch (7), and then the weight hanger (13) is suspended and the appropriate number weights (14) put on it. They give a static loading torque by twisting the shafts, which is measured indirectly using the torsion angle indicator (6). When the load has been applied, the two halves of the clutch (7) are firmly fixed against each other with the bolts. Then, the lock-pin is removed to close the safety cover. During the run, this loading torque "circulates" between the gears. In the back-to-back solution the motor (11) must overcome only the friction between gears, rolling bearings, and some minor components of friction (friction against seals, internal friction in the oil). Thus, the whole design is very simple and compact.

An AC squirrel-cage motor (11) of the nominal rotational speed of 3000 rpm is used to drive the rig. It is controlled by the frequency converter, which enables to change the rotational speed within a wide range.

In the gear scuffing tests the test gears are dip lubricated. In the test chamber where the test gears are located, there are heaters (15) to heat up the lubricating oil. The thermocouple (1), with the measuring point inserted in the lubricating oil, is to measure the oil temperature. A PID regulator is used to protect against overheating of the lubricating oil.

The motor (11) of the machine is automatically stopped when the preset time elapses. The required time is set on the controller panel. Additionally, the operator can read out the number of motor revolutions to confirm the correct duration of the run. The number of motor revolutions is displayed on the controller panel (Figure 8) connected to the speed transducer.

In the T-12U machine, the friction torque can be measured indirectly by measurement of the motor current load, which can be assumed to be proportional to the friction torque.

The test rig has a special support on the side cover of the test chamber (5) for mounting vibration transducers (accelerometers) to enable the operator to monitor the level of vibrations along different axes. However, now there is no possibility to automatically stop the motor when the vibration level is very high. This feature (together with other features like direct measurement of the friction torque) will be included in a new test rig, denoted as T-12UF, being developed at present.

Additional equipment includes a mass comparator for a very precise determination of the mass loss (wear) of the wheel.

Test Materials

The gears coated with the low-friction a-C:H:W coating (trade name: WC/C) of DLC type and composite low-friction MoS_2/Ti coating (trade name: MoST) were tested. All material combinations were tested: coating-coating (both gears coated), coating-steel, steel-coating, and steel-steel for reference (both gears without the coating). In all cases, mineral, automotive gear oil of API GL-5 performance level and of SAE 80W-90-viscosity grade was used for lubrication.

Statistical Analysis

To check statistical differences between the results obtained (FLS values), the uncertainty of measurement was assessed for the both developed test methods. This was done according to the procedures specified in the document EA-4/16 G: 2003, which are binding in the accredited laboratories meeting the requirements of ISO/IEC 17025:2005.

Once the uncertainty of measurement has been calculated, the test result "y" and the uncertainty of measurement "U" should be reported as "y ± U."

As a normal practice, the uncertainty of measurement is given in relation to the average value of the measurement. For example, in the case of the gear scuffing shock tests, the respective formula derived by the authors is expressed as follows:

$$U = 0.45 + 0.06 * FLS \qquad (1)$$

where:

U - uncertainty of measurement,

FLS - failure load stage.

According to ILAC-G8:03/2009, if the uncertainty intervals expressed by U do not overlap each other, one can say that the compared results are statistically different.

RESULTS AND DISCUSSION

Gear Scuffing EP Test for Coatings

Material Combinations With The A-C:H:W Coating

Failure load stages (FLS) obtained for the tested material combinations with the a-C:H:W coating are presented in Figure 10. The coated gear is dark grey coloured, and the uncoated one is light grey.

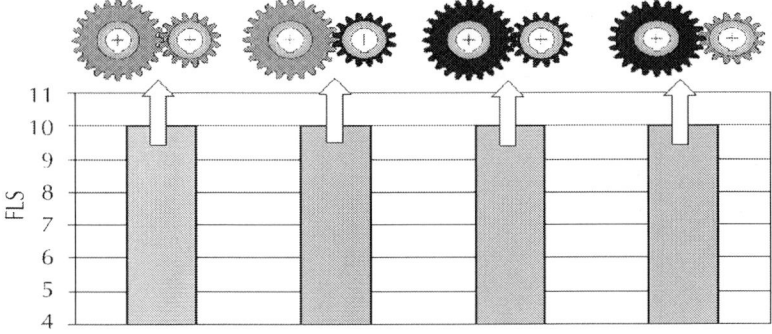

Figure 10: Failure load stages (FLS) obtained using the Gear Scuffing EP Test for Coatings for the tested material combinations with the a-C:H:W coating.

Figure 10 shows that the Gear Scuffing EP Test for Coatings is unable to differentiate between the tested material combinations from the point of view of the main criterion - FLS. All the FLS values exceed the maximum load stage, i.e. 10th. Thus, the additional criteria of failure assessment, related to the wear of the pinion after runs at particular load stages, were taken into account - Table 3. The table presents the symbolic modes of the wear of the test pinion at particular load stages for the tested material combinations with the a-C:H:W coating, and the mode of wear that appeared most often on the pinion teeth was considered. Below are the symbols of the wear modes, the total area of failures on the pinion (A_p) are given. The used symbols of wear were presented earlier in Table 2.

Table 3: Modes of the wear of the test pinion at particular load stages for the tested material combinations with the a-C:H:W coating, together with the total area of failures on the pinion (A_p); Gear Scuffing EP Test for Coatings

Load stage				
4	$A_p \approx 0$	$A_p \approx 0$	$A_p \approx 0$	$A_p \approx 0$
5	$A_p \approx 0$	$A_p \approx 0$	$A_p \approx 0$	$A_p \approx 0$
6	$A_p \approx 0$	$A_p \approx 0$	$A_p \approx 0$	$A_p \approx 0$
7	$A_p \approx 0$	$A_p \approx 0$	$A_p \approx 0$	$A_p \approx 0$
8	$A_p \approx 0$	$A_p \approx 0$	$A_p \approx 0$	$A_p \approx 0$
9	$A_p \approx 0$	$A_p \approx 0$	$A_p \approx 0$	$A_p \approx 0$
10	$A_p \approx 0$	$A_p \approx 0$	$A_p \approx 0$	$A_p \approx 0$

As can be observed in Table 3 for the tested material combinations, the three modes of wear that appear most often on the pinion teeth are scratches, polishing, and scoring. The uncoated pinion undergoes the process of polishing through the rubbing by the hard a-C:H:W coating deposited on the meshing wheel. Similar action was observed on the uncoated wheel meshing the coated pinion (results not shown here). The role of such polishing is to be explained in further experiments planned by the authors.

To sum up this part of the experiment, the Gear Scuffing EP Test for Coatings gives minor differences between the tested material combinations with the a-C:H:W coating, observed only when the pinion is uncoated and the wheel is coated. From the point of view of the practical applications of the a-C:H:W coating in gears, the situation when the both gears are coated seems to be better than in the case of one of the gears uncoated, because it is exposed to the abrasive action of the meshing coated gear, which results in polishing and scoring.

Material Combinations with the Mos$_2$/Ti Coating

Failure load stages (FLS) obtained for the tested material combinations with the MoS$_2$/Ti coating are presented in Figure 11. The coated gear is dark grey coloured, and the uncoated one is light grey.

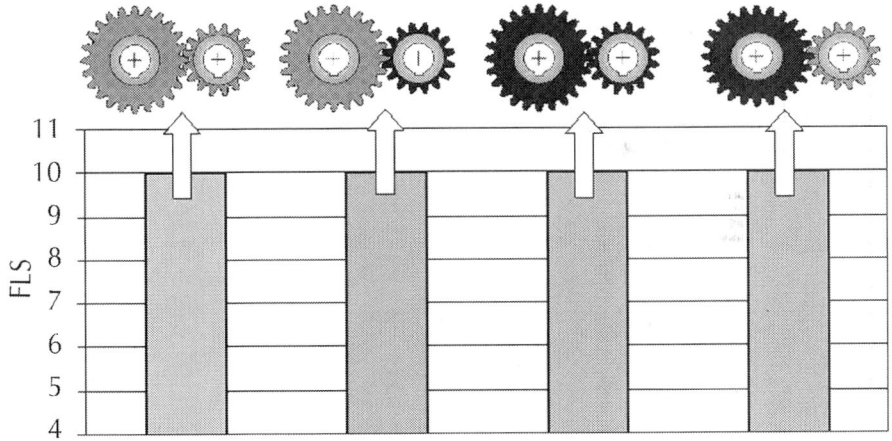

Combination of tested materials

Figure 11: Failure load stages (FLS) obtained using the Gear Scuffing EP Test for Coatings for the tested material combinations with the MoS$_2$/Ti coating.

Figure 11 shows that the Gear Scuffing EP Test for Coatings is unable to differentiate between the tested material combinations from the point of view of the main criterion - FLS. As in the case of testing the a-C:H:W coating, all the FLS values exceed the maximum load stage, i.e. 10th. Thus, the additional criteria of failure assessment, related to the wear of the pinion after runs at particular load stages, were taken into account - Table 4. The table presents the symbolic modes of the wear of the test pinion at particular load stages for the tested material combinations with the MoS_2/Ti coating, which is the mode of wear that appeared most often on the pinion teeth was considered. Below are the symbols of the wear modes, and the total area of failures on the pinion (A_p) are given. The used symbols of wear were presented earlier in Table 2.

Table 4: Modes of the wear of the test pinion at particular load stages for the tested material combinations with the MoS_2/Ti coating, together with the total area of failures on the pinion (A_p); Gear Scuffing EP Test for Coatings

Load stage				
4	$A_p \approx 0$	$A_p \approx 0$	$A_p \approx 0$	$A_p \approx 0$
5	$A_p \approx 0$	$A_p \approx 0$	$A_p \approx 0$	$A_p \approx 0$
6	$A_p \approx 0$	$A_p \approx 0$	$A_p \approx 0$	$A_p = 5$ m^{m}2
7	$A_p \approx 0$	$A_p \approx 0$	$A_p \approx 0$	$A_p = 10$ m^{m}2
8	$A_p \approx 0$	$A_p \approx 0$	$A_p \approx 0$	$A_p = 10$ m^{m}2
9	$A_p \approx 0$	$A_p \approx 0$	$A_p \approx 0$	$A_p = 10$ m^{m}2
10	$A_p \approx 0$	$A_p \approx 0$	$A_p \approx 0$	$A_p = 10$ m^{m}2

As can be observed in Table 4 for the tested material combinations, the two modes of wear that appear most often on the pinion teeth are scratches and scoring. In the material combination of the uncoated pinion meshing the coated wheel, the pinion bears the mark of scoring caused by the rubbing by the hard coating deposited on the meshing wheel.

Thus, the Gear Scuffing EP Test for Coatings gives minor differences between the tested material combinations with the MoS_2/Ti coating, observed only when the pinion is uncoated and the wheel is coated. From the point of view of the practical applications of the MoS_2/Ti coating in gears, the situation when the both gears are coated seems to be better than in the case of one of the gears uncoated as it is exposed to the abrasive action of the meshing coated gear, which results in scoring. However, when one of the gears needs to remain uncoated, using the a-C:H:W coating is more preferable than MoS_2/Ti, because a-C:H:W causes less wear of the uncoated gear.

Gear Scuffing Shock Test for Coatings

Material Combinations with the A-C:H:W Coating

Failure load stages (FLS) obtained for the tested material combinations with the a-C:H:W coating are presented in Figure 12. The coated gear is dark grey coloured, and the uncoated one is light grey. The assessed uncertainties of measurement for each result obtained are also shown in the Figure.

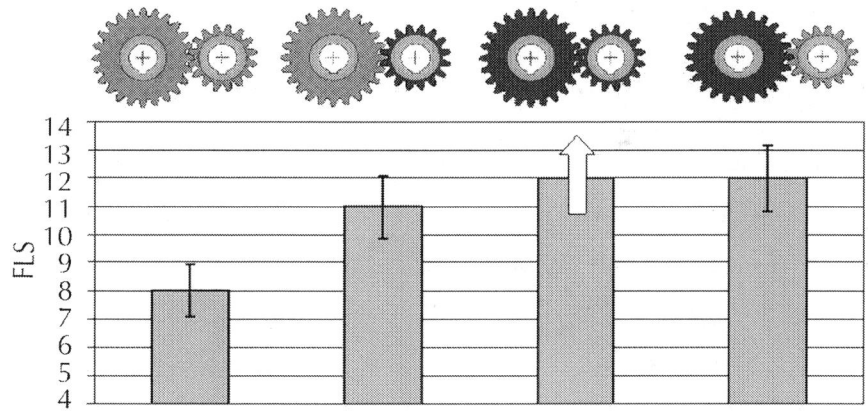

Combination of tested materials

Figure 12: Failure load stages (FLS) obtained using the Gear Scuffing Shock Test for Coatings for the tested material combinations with the a-C:H:W coating.

Figure 12 shows that the Gear Scuffing Shock Test for Coatings makes it possible to differentiate between the tested material combinations. The best resistance to scuffing (highest FLS) is observed when both gears are coated.

Under "shock" conditions, when the pinion is uncoated and the wheel is coated with the a-C:H:W coating, the resistance to scuffing is slightly higher than in the case when the pinion is coated and the wheel is uncoated. Hypothetically, there is a transfer of graphite (solid lubricant) from the a-C:H:W coated gear to the teeth of the uncoated one, which is more effective for the wheel coated than in the opposite situation, because the area of the coated steel surface of the wheel (larger gear with 24 teeth) is greater than in the case the coating is deposited on the pinion (small gear having only 16 teeth). However, one must have in mind that the difference in the scuffing resistance of the two material combinations is not statistically significant, because the measurement uncertainties overlap each other.

In comparison to the case of the both gears uncoated, when the a-C:H:W coating is deposited on one or two gears, much higher resistance to scuffing is observed. This is a result of a high surface energy for metals (here for steel) promoting adhesive bonding in the steel-steel contact, and smaller affinity in the different materials than when both of them are identical (i.e. steel-steel), which protects the surface from adhesive bonding. Yet another phenomenon can be attributed to it. When one of the mating materials (coating) is much harder than the other one (steel), or when two very hard materials are in contact (coating-coating) there is a reduction in the tendency to adhesive bonding, hence scuffing.

The additional criteria of failure assessment, related to the wear of the pinion after runs at particular load stages, were also taken into account - Table 5. The table presents the symbolic modes of the wear of the test pinion at particular load stages for the tested material combinations with the a-C:H:W coating, which is the mode of wear that appeared most often on the pinion teeth was taken into account. The photographs of the most often appearing mode of wear of the pinion at the highest load stage are shown also shown in the table. Red-shadowed cells in the table denote the failure load stage (FLS). Below are given the symbols of the wear modes, the total area of failures on the pinion (A_p), and wear of wheel (W_w). The used symbols of wear were presented earlier in Table 2.

As can be observed in Table 5 for the tested material combinations, the three modes of wear that appear most often on the pinion teeth are scratches, scuffing, and scoring. When one or both gears are a-C:H:W-coated, only scratches and scoring predominate on the pinion teeth.

What was observed also during the Gear Scuffing EP Test for Coatings, and what seems to by typical of "the action" of the a-C:H:W coating, the uncoated gear undergoes the process of polishing or scoring through the rubbing by the hard coating deposited on the meshing gear. The polishing on the wheel teeth flanks can be seen in Figure 13.

Figure 13: Photograph of the tooth flank of the uncoated wheel, polished by the a-C:H:W-coated pinion; Gear Scuffing Shock Test for Coatings.

Table 5: Modes of the wear of the test pinion at particular load stages for the tested material combinations with the a-C:H:W coating, together with the total area of failures on the pinion (A_p), and wear of wheel (W_w), obtained in the Gear Scuffing Shock Test for Coatings; red-shadowed cells - the failure load stage (FLS)

Load stage				
7	$A_p = 26$ mm² $W_w = 1$ mg			

8	$A_p = 703$ mm² W_w - not measured			
9				
10		$A_p \approx 0$ $W_w = 180$ mg		
11		$A_p \approx 0$ $W_w = 338$ mg	$A_p = 5$ mm² $W_w = 76$ mg	$A_p \approx 0$ $W_w = 2$ mg
12			$A_p = 6$ mm² $W_w = 145$ mg	$A_p = 318$ mm² $W_w = 3$ mg

Scuffing is observed only when both gears are uncoated - Table 5. This is one of the most dangerous modes of gear wear. As mentioned earlier, scuffing marks occur as single, fine marks or strips, or areas covering a part or all of the flank width. They appear as dull areas with the roughness much greater than the original criss-cross-grinding pattern shown in Figure 6. In this case, the grinding pattern is no longer visible.

To sum up this part of the experiment, the Gear Scuffing Shock Test for Coatings gives a much better resolution than the Gear Scuffing EP Test. However, one needs to have in mind that the cost of the former is about four-times higher than in the case of the latter, because the "shock tests" require more test gears to be used. From the point of view of the practical applications of the a-C:H:W coating in gears, the situation when the both gears are coated seems to be better than in the case of one of the gears uncoated, because it is exposed to the abrasive action of the meshing coated gear, which results in polishing or scoring. This positively verifies the observations taken during performing the Gear Scuffing EP Test for Coatings.

Material Combinations with the Mos₂/Ti Coating

Failure load stages (FLS) obtained for the tested material combinations with the MoS_2/Ti coating are presented in Figure 14.

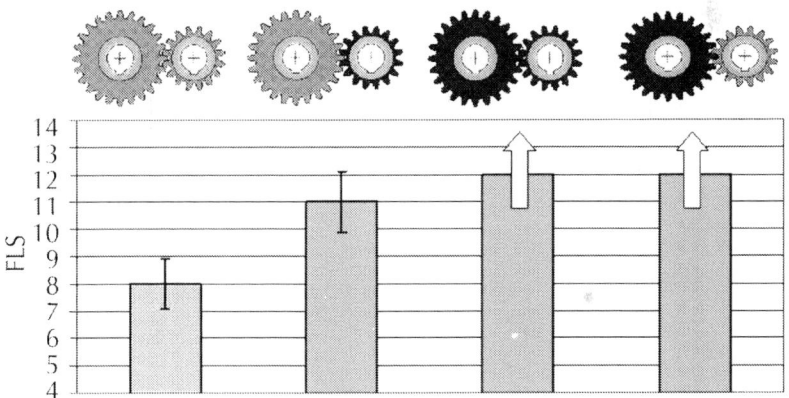

Combination of tested materials

Figure 14: Failure load stages (FLS) obtained using the Gear Scuffing Shock Test for Coatings for the tested material combinations with the MoS_2/Ti coating.

Figure 14 shows that the best resistance to scuffing (highest FLS) is observed when both gears are coated with the MoS_2/Ti coating, or when the uncoated pinion meshes the coated wheel.

As in the case of the a-C:H:W coating, when only the wheel is MoS_2/Ti-coated, under "shock" conditions the resistance to scuffing is higher than in the situation when only the pinion coated. Hypothetically, there is a transfer of MoS_2 (solid lubricant) from the teeth of the coated gear to the uncoated one. The transfer is more effective in the case of the MoS$_2$/Ti-coated wheel meshing the uncoated pinion than in the opposite situation, because the area of the coated steel surface of the larger gear (wheel) is greater than in the case when the coating is deposited on the small gear (pinion).

In comparison to the case when both gears are uncoated, a much higher resistance to scuffing is observed when the MoS$_2$/Ti coating is deposited on one or two gears. The respective mechanisms of this behaviour were described earlier.

Table 6 presents the symbolic modes of wear of the test pinion at particular load stages for the tested material combinations with the MoS$_2$/Ti coating. As in the case of the a-C:H:W coating, the mode of wear that appeared most often on the pinion teeth was taken into account. The photographs of the most often appearing mode of wear of the pinion at the highest load stage are also shown in the table. Red-shadowed cells in the table denote the failure load stage (FLS). Below are given the symbols of the wear modes, the total area of failures on the pinion (A$_p$) and wear of wheel (W$_w$). The symbols of wear were presented earlier in Table 2.

As can be observed in Table 6 for the tested material combinations, the two modes of wear that appear most often on the pinion teeth are scuffing and scoring. When one or both gears are coated, only scoring predominates on the pinion teeth.

When the pinion is coated and the wheel is uncoated, and when both the gears are coated, identical results were obtained for the two investigated coatings - FLS values are respectively 11 and higher than 12 (Figures 12 and 14). Therefore, the main criterion of assessment of the resistance to scuffing (FLS) makes it impossible to differentiate between these two situations. Under these circumstances, the analysis of additional criteria of failure assessment, related to the modes of wear at particular load stages, like in the previous cases can give additional, valuable information. In the case of the material combinations with the a-C:H:W coating, the predominating mode of wear of the pinion were only scratches. For the material combinations with the MoS$_2$/Ti coating, the pinion wear was much more sever, and scoring instead of scratches could be met most often. Thus, the a-C:H:W coating provides better protection against severe wear than MoS$_2$/Ti, especially when it is deposited on the both gears. This positively verifies the observations taken during performing the Gear Scuffing EP Test for Coatings.

Table 6: Modes of wear of the test pinion at particular load stages for the tested material combinations with the MoS_2/Ti coating, together with the total area of failures on the pinion (A_p), and wear of wheel (W_w), obtained in the Gear Scuffing Shock Test for Coatings; red-shadowed cells - the failure load stage (FLS)

Load stage				
7	A_p = 26 mm² W_w = 1 mg			
8	A_p = 703 mm² W_w - not measured			
9				
10		A_p = 32 mm² W_w = 11 mg		
11		A_p = 109 mm² W_w = 25 mg	$A_p \approx 0$ W_w = 16 mg	$A_p \approx 0$ W_w = 9 mg
12			$A_p \approx 0$ W_w = 16 mg	$A_p \approx 0$ W_w = 9 mg

SUMMARY AND CONCLUSIONS

The authors have developed unique test methods, being the subjects of this chapter. They are called the "Gear Scuffing EP Test for Coatings" and "Gear Scuffing Shock Test for Coatings."

The analysis of the values of the failure load stage (FLS), reflecting the resistance to scuffing, shows that the developed Gear Scuffing EP Test for Coatings has too little resolution to differentiate between the tested material combinations - coating-coating (both gears coated), coating-steel, steel-coating, and also steel-steel (both gears without a coating). Additional criteria of failure assessment need to be employed to reveal minor differences between the tested material combinations observed only when the pinion is uncoated and the wheel is coated.

In comparison, Gear Scuffing Shock Test for Coatings makes it generally possible to differentiate between the tested material combinations from the point of view of the main criterion of assessment of the gear resistance to scuffing, i.e. FLS. Thus, this test method has a sufficient resolution. However, as in the case of the Gear Scuffing EP Test for Coatings, apart from the analysis of only FLS values, analysis of the additional criteria of failure assessment related to predominating modes of wear at particular load stages is recommended and may give additional, valuable information. For the two coatings tested (a-C:H:W and MoS_2/Ti), the best resistance to scuffing/scoring (FLS > 12) is observed when both gears are coated; however, the a-C:H:W coating gives a slightly better protection against severe wear than MoS_2/Ti - only scratches instead of scoring are observed for a-C:H:W.

Although the Gear Scuffing Shock Test for Coatings gives a much better resolution than the Gear Scuffing FP Test, one needs to have in mind that the cost of the former is about four-times higher than in the case of the latter, because the "shock tests" require more test gears to be used.

In the both tests, when one or both gears are coated, three modes of wear occur most often on the pinion teeth - polishing, scratches, or scoring. Scuffing is observed only when the two gears are uncoated.

The following conclusions can be drawn:

- The developed Gear Scuffing Shock Test for Coatings has been successfully verified by the testing of thin, hard coatings deposited on the gears; therefore, it can be implemented in the laboratories of the R&D centres devoted to surface engineering and the engineering of advanced materials intended for modern toothed gears, having in mind that this test is rather expensive.

- If the coating is intended for application on gears, from the point of view of the highest achievable resistance to scuffing/scoring, it is recommended that both meshing gears are a-C:H:W-coated.

- Although the T-12U gear test rig has been effectively employed in the performed research, it is suggested that its research capacities should be extended by the measurement and data acquisition of the friction torque, which will make it possible to investigate, as postulated by gear transmissions manufacturers, the possibility of the reduction of friction between the meshing teeth by the application of a low-friction coating. At present, a new version of the T-12U test rig, denoted as T-12UF, is being developed within the framework of the Strategic Programme executed at ITeE-PIB in Radom, and the planned deadline of this work is in 2013.

- Both the differentiation between the tested objects (lubricating oils, material combinations) and the predictability of gear failures in real applications (transmissions) are important when assessing gear tests. This is why the authors plan to verify the results obtained by application of coated gears in transmissions (speed reducers) of different devices manufactured by one of the Polish producers. What is more, at present another test rig - a back-to-back bevel gear test rig, denoted as T-30 - is being developed in the Tribology Department of ITeE-PIB in Radom with the deadline in 2012. The reason is that until now widely used test devices and methods have allowed researchers to perform runs on only spur gears having the tooth geometry significantly different than the geometry of bevel gears. The new tribotester will allow researchers to better predict the failures of bevel gears.

ACKNOWLEDGEMENTS

Scientific work was financed:

- From the means of the Minister of Science and Higher Education, executed within the Strategic Programme "Innovative Systems of Technical Support for Sustainable Development of the Country's Economy" within Innovative Economy Operational Programme.

- By the National Centre for Research and Development (NCBiR) within the scope of the R&D project No. N R03 0019 06.

The authors wish to express their thanks also to Dr. Maksim Antonov from Tallinn University (Estonia) for his support with the gear scuffing tests and helpful discussions, within the framework of Marie Curie RTN (6th EU FP); Contract No MRTN-CT-2006-035589.

REFERENCES

1.   Pytko S., Sroda P. Classification and Evaluation of Machines for Investigation of Materials for Production of Gear Wheels. ZEM 1975; 1 39-58 (*in Polish*).

2.   Tomaszewski J., Drewniak J., editors. Scuffing of Gears. Gliwice: CMG KOMAG; 2007 (*in Polish*).

3.   ARTEC Machine Systems. Failure by scuffing due to poor distribution of load. http://www.artec-machine.com/images/failure_due_to_poor_distribution_of_load_1.pdf (accessed 22 September 2012).

4.   Kratz S.H., Wedeven L.D., Black W.F., Carlisle D.J., Wedeven G.G. Simulation of Space Shuttle Gear Performance. In: proceedings of the III World Tribology Congress, 2005, Washington, USA.

5.   Kuo W.F., Chiou Y.C., Lee R.T. A Study on Lubrication Mechanism and Wear Scar in Sliding Circular Contacts. Wear 1996;201 217-226.

6.   Ku P.M., editor. Interdisciplinary Approach to Friction and Wear. Washington D.C.: Southwest Research Institute; 1968.

7.   Makowska M., Matuszewska A., Gradkowski M. Migration of Active Elements from a Lubricant to the Material of the Friction Pair. Tribologia 2011;4 163-176 (*in Polish*).

8.   Wachal A. Analysis of Boundary Layer Estimating Criteria in Lubricating Oils Investigations. ZEM 1983;3 325-332.

9.   Höhn B.-R., Oster P., Schedl U. Pitting Load Capacity Test on The FZG Gear Test Rig with Load-Spectra and One-Stage Investigations. Tribotest journal 1999;5-4 417-430.

10.   Szczerek M., Tuszynski W. A Method for Testing Lubricants under Conditions of Scuffing. Part I. Presentation of the Method. Tribotest journal 2002;8-4 273-284.

11. Piekoszewski W., Szczerek M., Tuszynski W. The Action of Lubricants under Extreme Pressure Conditions in a Modified Four-Ball Tester. Wear 2001;249 188-193.

12. Tuszynski W., Michalczewski R., Piekoszewski W., Szczerek M. Effect of Ageing Automotive Gear Oils on Scuffing and Pitting. Tribology International 2008;41 875-888.

13. Bisht R.P.S., Singhal S. A Laboratory Technique for the Evaluation of Automotive Gear Oils of API GL-4 Level. Tribotest journal 1999;6-1 69-77.

14. Method to Assess the Scuffing Load Capacity of Lubricants with High EP Performance Using an FZG Gear Test Rig. FVA Information Sheet No. 243 Status June 2000.

15. Höhn B.-R., Michaelis K., Eberspächer C., Schlenk L. A Scuffing Load Capacity Test with the FZG Gear Test Rig for Gear Lubricants with High EP Performance. Tribotest journal 1999;5-4 383-390.

16. Michaelis K., Höhn B.-R., Graswald C. Scuffing Tests for API GL-1 to GL-5 Gear Lubricants. In: proceedings of the 13th International Colloquium Tribology, 2002, Ostfildern, Germany.

17. Michaelis K., Höhn B.-R., Oster P. Influence of Lubricant on Gear Failures - Test Methods and Application to Gearboxes in Practice. Tribotest journal 2004;11-1 43-56.

18. Höhn B.-R., Oster P., Tobie T., Michaelis K. Test Methods for Gear Lubricants. Goriva i maziva 2008;47-2 141-152.

19. Tuszynski W. Performance Classification of Automotive Gear Oils Using the Gear Scuffing Shock Test. Tribologia 2009;2 259-274 (in Polish).

20. Kalin M., Vižintin J. The Tribological Performance of DLC-Coated Gears Lubricated with Biodegradable Oil in Various Pinion/Gear Material Combinations. Wear 2005;259 1270-1280.

21. Martins R.C., Moura Paulo S., Seabra J.O. MoS2/Ti Low-Friction Coating for Gears. Tribology International 2006;39 1686-1697.

22. Martins R., Amaro R., Seabra J. Influence of Low Friction Coatings on The Scuffing Load Capacity and Efficiency of Gears. Tribology International 2008;41 234-243.

23. Szczerek M., Michalczewski R., Piekoszewski W. The Problems of Application of PVD/CVD Thin Hard Coatings for Heavy-

Loaded Machine Components. In: proceedings of the ASME/STLE International Joint Tribology Conference, 2007, San Diego, USA.

24.   Michalczewski R., Piekoszewski W., Szczerek M., Tuszynski W. The Lubricant-Coating Interaction in Rolling and Sliding Contacts. Tribology International 42;2009 554-560.

25.   Michalczewski R., Piekoszewski W., Szczerek M., Tuszynski W. Scuffing Resistance of DLC Coated Gears Lubricated with Ecological Oil. Estonian Journal of Engineering 2009;15-4 367-373.

26.   CEC L-07-95: Load Carrying Capacity Test for Transmission Lubricants (FZG Test Rig).

Chapter 4

Artificial Slip Surface: Potential Application in Lubricated MEMS

M. Tauviqirrahman[1,2], R. Ismail[1,2], J. Jamari[1], and D.J. Schipper[2]

[1]Laboratory for Engineering Design and Tribology, Mechanical Engineering Department, University of Diponegoro, Jl. Prof. H. Sudharto, Kampus UNDIP Tembalang, Semarang, Indonesia

[2]Laboratory for Surface Technology and Tribology, University of Twente Drienerlolaan, Enschede, The Netherlands

INTRODUCTION

Background

For the last years, there has been a tremendous effort towards the development of Micro-Electro-Mechanical System (MEMS) for a wide variety of applications in aerospace, automotive, biomedical, computer, agricultural industries, electronic instrumentation, industrial process control, biotechnology, office equipment, and telecommunications. MEMS devices integrate chemical, physical, and even biological processes in micro-scale technology packages.

Stiction (a subtraction of 'static friction') in micro-system technology has been a problem ever since the advent of surface micromachining in the eighties of the last century. As the overall size of the machine is reduced, the capillary and surface tension force of liquid become large, which induce stiction rendering the devices to fail or malfunction. In particular, stiction forces created between moving parts that come into contact with one another, either intentionally or accidentally, during operation are a common problem with micro-mechanical devices. Stiction-type failures occur when the interfacial attraction forces exceed restoring forces. Consequently, the surfaces of these parts either temporarily or permanently adhere to each other, causing device malfunction or failure.

Several approaches to address the stiction between two opposing surfaces have been presented in the various literatures [1-4]. The basic approaches to prevent stiction include increasing surface roughness (topography) and/or lowering solid surface energy by coating with low surface energy materials. This includes self-assembled molecular (SAM) coatings, hermetic packaging and the use of reactive materials in the package [5].

Other attractive technique to tackle the stiction problem is by inserting a lubricant into the region around the interacting devices to reduce the chance of stiction-type failures. As is well-known, many MEMS devices include moving (sliding/rolling) surfaces and thus it is necessary to apply a lubricant between the contacting surfaces to reduce friction and wear. However, a significant barrier to the development of MEMS lubrication is the problem of achieving effective tribological performance of their moving parts. This is because the lubricant behavior is different at micro-scale compared to macro-scale. At the macroscopic level, it is well accepted that the boundary condition for a viscous fluid at a solid wall is no-slip, i.e. the fluid velocity matches the velocity of the solid boundary. While the no-slip boundary condition has been proven experimentally to be accurate for a number of macroscopic flows, it remains an assumption that is not based on physicals principles. At micro-scale level, certain phenomena must be taken into account when analyzing liquid flows such as a slip condition at solid wall boundaries.

As a consequence of the MEMS technology revolutionary application to many areas, it is possible for scientists to observe the boundary slip on micro/nano-meter scale. A variety of techniques are now available that are capable of probing lubricant flow on micro-scales and are therefore suitable for the investigation of boundary conditions. There are three techniques so far for detecting the boundary slip: nano-particle image velocimetry (NPIV) [6], atomic force microscope (AFM) [7-9] and surface force apparatus (SFA) [10]. The NPIV technique is a direct observation method with a measurement precision depending on the size of the nano-particles but with poor moderate accuracy. The AFM and SFA are indirect observation techniques based on the assumption that boundary slip takes place precisely on the interface of liquid and solid. These methods need a high accuracy boundary slip model to infer the slip velocity. Boundary slip has been observed not only for a hydrophobic surface [6, 7, 10] but also for a hydrophilic surface [8, 9]. Therefore, the slip evidence has been generally accepted and for certain cases the no-slip boundary condition is not valid.

There is a large body of literature dealing with the analysis of lubricant slip flow based on the analytical and numerical solution of molecular dynamic simulations [11, 12], Lattice-Boltzman [13, 14], and the Reynolds equation [15-23]. The accurate description of slip at the wall is very difficult and still remains a subject of intensive research. The so-called Navier slip model and the critical shear stress model are usually used to describe a boundary slip. In fact, nearly two hundred years ago Navier [24] proposed a general boundary condition that incorporates the possibility of fluid slip at a solid boundary. Navier's proposed boundary condition assumes that the velocity, u, at a solid surface is proportional to the shear stress at the surface. It reads: $u = b \, (du/dz)$ where b is the slip length and du/dz is the shear rate. The slip length b, which is defined as the distance beyond the solid/liquid interface at which the liquid velocity extrapolates to the velocity of the solid, is used to quantify a boundary slip. If $b = 0$ then the generally assumed no-slip boundary condition is obtained. If $b =$ finite, fluid slip occurs at the wall, but its effect depends upon the length scale of the flow. The Navier-slip boundary condition is the most widely used boundary condition with the methods based on the solution of continuum equations.

In micro-scales such as MEMS, the boundary condition will play a very important role in determining the lubricant flow behavior. Control of the boundary condition will allow a degree of control over the hydrodynamic pressure in confined systems and be important in lubricated-MEMS. To prevent a stiction, in a controlled way, one is able to enhance, a hydrophobic/hydrophilic behavior of surfaces. If one surface is hydrophobic (slip) and the other is hydrophilic (no-slip) the sliding velocity or displacement between the surfaces is accommodated by shear at the hydrophobic surface (the lubricant is kept in the contact by the hydrophilic surface). In this way wear of the surfaces is prevented and the surfaces are able to move because stiction is prevented.

The slip situation, however, can be controlled to obtain a positive effect by surface technology. Coating and texturing technologies can be used to engineer large slip. In practice, a large slip can be made using super-hydrophobic surfaces. Such surfaces can be manufactured by grafting or by deposition of hydrophobic compounds on the initial surface at a certain zone. Super-hydrophobic surfaces were originally inspired by the unique water-repellent properties of the lotus leaf. It is the combination of a very large contact angle and a low contact-angle hysteresis that defines a surface as super-hydrophobic. Implementing the slip property (hydrophobicity) on a surface in a wide range of application for the mechanical components is of great challenge by numerous authors recently. In published works [15-23], both experimentally and numerically, slip surface is able to reduce friction force at the contacting surfaces and finally reduce energy consumption, increase component's life-time and reduce economic and environmental costs.

Problem Statement

In classical liquid lubrication it is assumed that surfaces are fully wetted and no-slip occurs between the fluid and the solid boundary. In MEMS, this wetting is actually an unwanted process because it can encourage the occurrence of stiction and as a result micro-parts can not be moved [25]. It is expected that slip can reduce the friction and improve the load support. However, with respect to the engineered slip pattern, the choice of slip zone on a certain surface must be taken carefully in relation to such tribological performances. In other words, an inappropriate slip zone pattern on a certain surface or the election of inappropriate surface containing a slip situation may lead to the deterioration of the lubrication performance. How to control the boundary slip in the application of a lubricated-MEMS is one of the challenging tasks in the future. This chapter will explore the provision of a new lubrication model based on the continuum approach for moving parts in MEMS in order to improve the tribological performance of lubricated contacts. In MEMS, by lubrication, low friction force and high load support are the goals which want to be achieved. The artificial slip surface will be introduced as one of the solutions to improve the lubrication performance of MEMS so as MEMS with a longer life-time can be obtained. The term "artificial slip surface" is used to address a non-homogeneous engineered slip/no-slip pattern, i.e. a surface consisting of a slip zone and a no-slip zone.

RESEARCH METHODS

Full film lubrication of lubricated contact is often described by the Reynolds theory [26]. According to the classical Reynolds theory, no-slip boundary is assumed and the convergent geometrical wedge is one of the most important conditions to generate hydrodynamic pressure. Therefore, the lubrication model for lubricated MEMS will be an extension of the classical lubrication theory. This means that modeling the lubricant through very narrow gap, normally modeled by assuming no-slip at the boundaries will be modified by introducing a boundary slip.

Modified Reynolds Equation

The classical Reynolds equation that is valid under no-slip condition can be generalized for taking into account slip conditions. It is then possible, for any film height distributions, to calculate the pressure distribution and the shear rate profile. The model of lubrication presented here is based on the fact that slip of the lubricant will exist at the interface of a lubricated sliding contact. Thus, a boundary slip is employed both on the moving and stationary surface, see Figure 1. The proposed lubrication model with slip leads to a modified Reynolds equation as presented in Eq. (1).

$$\frac{\partial}{\partial x}\left(\frac{h^3}{12\mu}\frac{h^2+4h\mu(\alpha_t+\alpha_b)+12\mu^2\alpha_t\alpha_b}{h(h+\mu(\alpha_t+\alpha_b))}\frac{\partial p}{\partial x}\right)=\frac{u_w}{2}\frac{\partial}{\partial x}\left(\frac{h^2+2h\mu\alpha_t}{h+\mu(\alpha_t+\alpha_b)}\right)-u_w\frac{\alpha_t\mu}{h+\mu(\alpha_t+\alpha_b)}\frac{\partial h}{\partial x}+\frac{h}{2\mu}\frac{\partial p}{\partial x}\frac{\partial h}{\partial x}\frac{h\alpha_t\mu+2\alpha_t\alpha_b\mu^2}{h+\mu(\alpha_t+\alpha_b)} \quad (1)$$

The physical meanings of the symbols in Eq. (1) are as follows: h the lubrication film thickness (gap) at location, p the lubrication film pressure, μ the lubricant viscosity, the slip coefficient (subscripts t and b denote the top (stationary) and bottom (moving) surface, respectively) and u_w the velocity of the moving surface. It can be seen in that if the slip coefficient is set to zero (no-slip condition), Eq. (1) reduces to the classical Reynolds equation. It should be noted that the product of multiplication of the slip coefficient by the viscosity, μ, is usually called as 'slip length' b.

Eq. (1) is derived by following the usual approach to deduce the Reynolds equation from the Navier-Stokes system by assuming classical assumptions except that boundary slip is applied both on the stationary surface and moving surface as depicted in Figure 1.

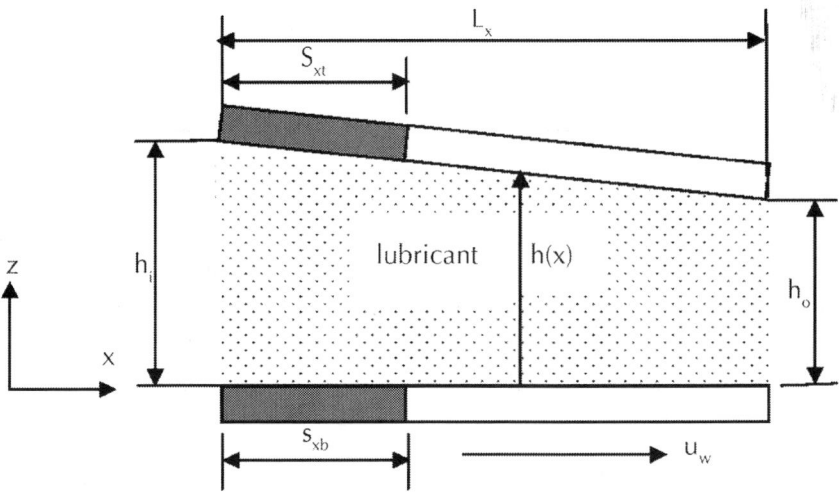

Figure 1: Schematic of a lubricated sliding contact with artificial slip surface applied both on the stationary and the moving surface. The boundary slip zones (S_{xt} and S_{xb}) are located at the leading edge of the contact. (Note: u_w is the sliding velocity, L_x is contact length, h_i and h_o are inlet and outlet film thickness, respectively, $h(x)$ is variable film thickness).

Eq. (1) can be derived by considering the equilibrium of an element of fluid.

$$\frac{\partial^2 u}{\partial z^2} = \frac{1}{\mu}\frac{\partial p}{\partial x} \tag{2}$$

where z lies along the direction across the thickness of the film, and u is the velocity field. To obtain the velocity profile, Eq. (2) can be integrated twice.

$$u = \frac{1}{2\mu}\frac{\partial p}{\partial x}z^2 + C_1 z + C_2 \tag{3}$$

C1 and C2 in this case are either constants or functions of x and can be solved by applying a boundary condition for u. The bottom and the top surfaces have the slip condition based on slip length model.

$$\left. \begin{array}{l} \text{at } z \;=\; 0, \; u = uw + \alpha_b\mu\dfrac{\partial u}{\partial z}\bigg|_{z=0} \\[2em] \text{at } z \;=\; h, \quad u = -\alpha_t\mu\dfrac{\partial u}{\partial z}\bigg|_{z=h} \end{array} \right\}$$

(4)

This gives

$$u = \frac{1}{2\mu}\frac{\partial p}{\partial x}z^2 - \left(\frac{u_w}{h+\mu(\alpha_t+\alpha_b)} + \frac{h}{2\mu}\frac{\partial p}{\partial x}\frac{h+2\alpha_t\mu}{h+\mu(\alpha_t+\alpha_b)} \right)z + u_w\frac{h+\alpha_t\mu}{h+\mu(\alpha_t+\alpha_b)}$$

(5)

$$-\frac{h}{2\mu}\frac{\partial p}{\partial x}\frac{\alpha_b\mu(h+2\alpha_t\mu)}{h+\mu(\alpha_t+\alpha_b)}$$

(6)

This velocity is used to compute the flow rate, q by integrating across the fluid film thickness, h. When q is differentiated to fulfill the continuity of flow, assuming μ is constant; this gives a modified Reynolds equation as stated in Eq. (1).

When full film lubrication is assumed, the entire load w is carried by the lubricant film and the calculation is simply an integration of the lubricant film pressure distribution over contact area, i.e.

$$w = \int_0^{L_x} p(x)\,dx$$

(7)

The friction force f generated by the lubrication system is due to the fluid viscous shear. It is calculated by integrating the shear stress over the surface area. These shear stresses are given by

$$\tau(x,z) = \left(\mu\frac{\partial u}{\partial z} \right)_{z=h}$$

(8)

The simulation results will be presented in dimensionless form, i.e.

$P = ph_0^2 / \mu L_x u_w$ for the dimensionless pressure, $W = wh_0^2 / (u_w \mu L_x^2)$ for the dimensionless load support (where w is the load per unit width), $F = fh_0/\mu u_w L_x$ for dimensionless friction force (where f is the friction force per unit width), and m = F / W for dimensionless friction coefficient. In the present study, the dimensionless slip length A is determined by normalizing the slip length b with the outlet film thickess h_0. For slip analysis in the following computations, the dimensionless slip length A varies from 3 to 300, which are reasonable values of the slip length based on the results published in literature [17, 18, 20].

Solution Method

The modified Reynolds equation, Eq. (1) is discretized over the flow using the finite volume method, and is solved using the tridiagonal matrix algorithm (TDMA), [27]. By employing the discretization scheme, the computed domain is divided into a number of control volumes using a grid with uniform mesh size. The grid independency is validated by various numbers of mesh sizes. An assumption is made that the boundary pressures are zero at both sides of the contact.

A numerical simulation is conducted to investigate the possible application so as a boundary slip can be beneficial to achieve a high load support and low friction force. In order to maximize the performance of lubrication, the boundary conditions (slip zones, S_{xt} and S_{xb}, see Figure 1) of the model are optimized through a parametric analysis. The object of optimization is to maximize the hydrodynamic load support. The load support satisfies two main functional purposes: (a) carry the applied external load, and (b) to minimize the contact between the opposing solids, and thus wear. The optimization analysis attempts to satisfy both functional requirements with a single design parameter, the slip zone.

The parametric analysis is performed using a developed computer code to investigate the effect of various slip parameters on the lubrication performances (load support, friction force, and friction coefficient). A parametric study is conducted with the variation of slip parameters (slip zone and slip length) over a large range of values considering different performance parameters. The design variables and the objective function are referred to as the optimization variables.

The design variables are independent quantities which are varied in order to achieve the optimum design. The objective function is the dependent variable that is maximized, i.e. the load support. In the present study, the design variables are slip zones as indicated in Figure 1. The algorithm used in the present study is depicted on Figure 2.

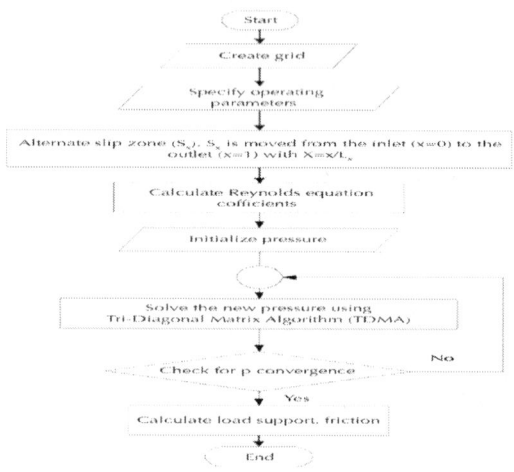

Figure 2: Flow chart for numerical method.

KEY RESULTS

The behavior of traditional (no-slip) hydrodynamic lubrication between the opposing surfaces can be estimated by a classical form of the Reynolds equation. The derivation of the classical Reynolds equation is based on the assumption of no-slip between the lubricant and the surfaces. In the classical Reynolds lubrication, the mechanism to generate a pressure is due to the convergent wedge effect. An artificial slip surface presented here is designed to be able to carry the external load during lubrication even if the wedge effect is not present. This situation is very beneficial in designing lubricated-MEMS which exhibits parallel gaps.

In this chapter, there are two main investigations. At first, the study is conducted in order to validate the developed numerical scheme. It assures that the numerical method used can be employed for solving other hydrodynamic characteristics. The no-slip case of lubricated contact is of main interest due to the availability of the analytical solution. Secondly, the study will be extended to explore the effect of the slip zone of the artificial slip surface on pressure, load support, friction force, and friction coefficient. The comparison between the modified sliding contact containing an artificial slip surface and the traditional one is conducted in order to describe the benefit of the use of an artificial slip pattern quantitatively.

No-Slip Condition

The modified Reynolds equation (Eq. (1)) is the governing equation for the fluid lubrication system containing a boundary slip. If the slip coefficient, , is set to zero, Eq. (1) reduces to the classical Reynolds equation. In this section, in order to validate the developed computer code containing a numerical scheme using finite volume method combined with tridiagonal matrix algorithm (TDMA), the classical Reynolds equation (no-slip condition) is solved numerically for calculating the pressure distribution, and finally the friction in a lubricated sliding contact as depicted in Figure 1 and Table 1, respectively. These results are compared with the analytical solution based on the work of Cameron [28] as follows:

$$p = \frac{6u_w \mu L_x}{h_0^2} \frac{K \frac{x}{L_x}\left(1 - \frac{x}{L_x}\right)}{(2+K)\left(1+K-K\frac{x}{L_x}\right)^2}$$

(9)

for the pressure distribution where $K = (h_i / h_o) - 1$, and

$$f = \frac{L_x \mu u_w}{h_o}\left(\frac{4\ln(1+K)}{K} - \frac{6}{(2+K)}\right)$$

(10)

for the friction force per unit width.

In Figure 3 the numerical results obtained with TDMA as well as analytical results for the dimensionless pressure distribution along the bottom wall of the contact are shown alongside those obtained from the Reynolds approximation. The wedge ratio h* of 2.2 was considered based on the fact that a maximum load support for a no-slip contact occurs when h* = 2.2 [28]. In the present study the wedge ratio h* is defined as the inlet film thickness over the outlet film thickness, hi/h0, and sometimes quoted as slope incline ratio in other literature. It is observed from Figure 3 that the maximum error is within 0.01% between the pressure obtained from the analytical solution and the numerical result.

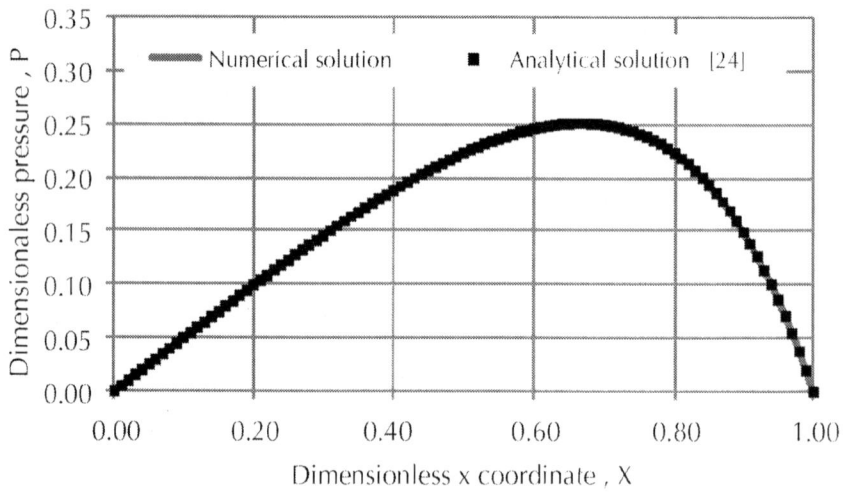

Figure 3: Normalized pressure distribution along the bottom wall of the linear wedge with no-slip boundary condition for an optimal wedge ratio h* of 2.2.

The comparison between the dimensionless friction force F obtained with the numerical prediction and the analytical solution are presented in Table 1. Like the result of the pressure distribution, the predicted dimensionless friction force shows very good agreement with the analytical solution. In general, the numerical solution of the classical Reynolds equation is matched well with the analytical solution. It assures that the numerical method used is valid and thus can be extended for analysing other hydrodynamic characteristics.

Table 1: The comparison between the analytical solution performed by Cameron [28] and the numerical simulation code

	Dimensionless friction force, F
Analytical solution [28]	0.77
Numerical prediction [present study]	0.77

Artificial Slip Surface

In MEMS, liquid lubrication has generally been omitted due to high hydrodynamic friction force that occurs in fluid film. Compared with a solid coating, stiction prevention using liquid lubrication is less practical. However, recent studies have demonstrated that it is possible for Newtonian liquids to slip along very smooth solid walls [20] and this result may make liquid lubricants for MEMS devices feasible. The main advantage of a liquid lubricant over a solid lubricant is that they generally produce no-contact shear stresses. Unfortunately, a stiction-type failure due to a large shear and capillary forces occurs. In [15, 16], a lubrication model of low load contacts was proposed to reduce such stiction or friction. The idea behind that work was how to use a lubricant that does not wet one of the solid surfaces. It was found that a half-wetted bearing generates a significant friction reduction compared to a traditional bearing.

In order to reduce stiction, two principal methods are available, chemical and physical modification of the surfaces. To generate wall slip, in the chemical approach, the chemical composition of the surface is altered. In the physical approach, the surfaces are roughened to decrease the effective contact area [29].

In practice, the slip zone of the artificial slip surface can be prepared from (super) hydrophobic surface which uses chemical properties as well as micro- and nano-structures in order to achieve a high level of friction force reduction. The main characteristic of (super) hydrophobic surfaces is the slip length. Extensive studies have confirmed that the chemical treatment of the surfaces generates a slip length in the order of 1 μm [30], while longer slip length up to 100 μm can be obtained through a combination of a hydrophobic surface with textured structure [20, 31, 32]. In the present study, it will be shown by the computational analysis that a longer slip length applied on the slip zone of the artificial slip surface leads to a greater friction force reduction in combination with an improved load support.

Beneficial Surface of Slip

Recently, the use of an engineered slip surface has become popular with respect to lubrication, since this type of surface enhancement would give a better tribological performance. The great challenge for an engineered slip surface from the perspective of a numerical simulation is choosing the optimal slip zone geometry with respect to the lubrication performance. Two engineered slip surface modes were used currently: homogeneous slip surface (i.e. slip applied over the whole surface) and artificial slip surface (i.e. surface consisting of slip zone and no-slip zone). It can be noted that term "artificial slip surface" was sometimes also called as heterogeneous slip/no-slip surface [17, 18] and mixed slip surface [19]. The first study to mention using a homogeneous slip was dedicated by Spikes [15, 16] who numerically studied the effect of slip profiles on friction. The author pointed out that by introducing the half-wetted bearing having a homogeneous slip boundary on one of the surfaces, a reduced friction can be obtained. Subsequently, an experimental study was published in [20] confirming the finding of [15, 16]. However, in addition to the friction reduction, it was shown that a homogeneous slip surface usually has a negative effect, i.e. the decrease in the load support. If the lubricated contact exhibits a perfect slip property, it was found that the fluid load support was only half of that without slip [115, 19, 21, 23]. Clearly, this is unwanted effect with respect to the lubrication. Therefore, to date, an artificial slip surface has become of great interest by some researchers [17-19, 21, 23] with the focus of how to balance the slip effect on the load support and friction.

The big question with respect to the tribological performance of lubricated-MEMS emerges in accordance with at which wall boundary slip must be applied, at the stationary surface, moving surface, or both of them. Besides that, the types of slip zone pattern become also great issue. Therefore, a series of simulations were conducted with such boundaries to find the best possibility of slip boundary application in terms of load support. Investigations were made for four kinds of slip boundaries to find the best boundary slip in terms of tribological performance, i.e. (1) slip applied on both the stationary and moving surfaces is referred as 'condition 1', (2) slip applied on the stationary surface is referred as 'condition 2', (3) slip applied on the moving surface is referred as 'condition 3', and (4) no-slip condition applied on the both of surfaces is referred as 'condition 4'. Here, a homogeneous slip surface is employed for all slip conditions.

Figure 4 presents the effect of the wedge ratio h^* on the dimensionless load support W. It is shown that the contact with homogeneous slip condition of 1, 2, and 3 have a negative effect, i.e. a reduced load support. The highest achievement of a load support W is obtained when the slip is applied on the stationary surface (condition 2). However, the value of the predicted load support is much lower than the conventional lubricated contact for all values of wedge ratio. It is only half of what the conventional Reynolds theory predicts for an optimal wedge ratio of a traditional slider contact. Fortunately, the direct trend of homogeneous slip to decrease the load support W is counterbalanced by the fact that such surface also reduces friction significantly. This is indicated in Figure 5 which shows the comparison between the dimensionless friction F for condition 2 and the condition 4 for the range of wedge ratio, h^*. The reduced friction as an advantageous effect in the lubrication can be explained by the fact that the boundary slip tends to reduce the wall shear rate at a prescribed film thickness, and then the wall shear stress and friction. Therefore, in the following design for the maximal lubrication performance, the lubricant has a no-slip boundary condition at the moving solid surface but can slip along the stationary surface. In the next section, the geometry of the boundary slip zone making an artificial slip surface will be investigated in order to achieve a higher load support as well as a low friction force.

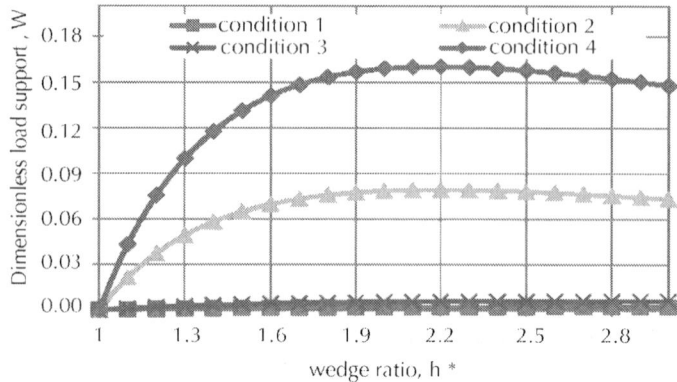

Figure 4: Dimensionless load support versus wedge ratio for several homogeneous boundary slip conditions. (Note: slip on stationary and moving surfaces (condition 1); slip on stationary surface (condition 2); slip on moving surface (condition 3); traditional no-slip (condition 4)). For the slip cases, the dimensionless slip length A of 20 is assumed.

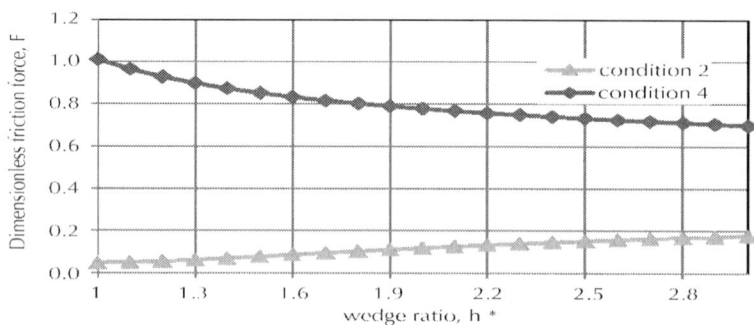

Figure 5: Dimensionless friction force versus wedge ratio at boundary condition in which slip applied on the stationary surface (condition 2) compared to the no-slip condition (condition 4).

The Optimum Slip Zone of the Artificial Slip Surface

This section is intended to investigate the optimum slip zone of the artificial slip surface for several values of wedge ratios h^* with respect to the load support.

In the following computations, as discussed in the previous section, for a high load support, it is considered that the artificial slip surface will take place on the stationary surface, whereas no-slip condition occurs on the moving surface. The parameter S_{xD} (in which $S_{xD} = S_x/L_x$) is introduced in order to completely define the dimensionless slip zone. A parametric analysis is conducted by varying the dimensionless slip zone S_{xD} from zero (i.e. no-slip boundary) to one (i.e. homogeneous boundary slip). The effect of slip zone geometry on the load support is presented in Figure 6. It can be shown that the load support has a maximum value when S_{xD} = 0.65 and h^* = 1 (i.e. parallel sliding surfaces). It should be noted that no hydrodynamic pressure can be built up in parallel sliding surfaces for traditional no-slip contact. From Figure 6, it is also shown that with the increase in wedge ratio h^*, a clear shift of the maximum load support toward to outlet can be observed in this work. In addition, the maximum load support decreases with the increase in h^*. A first conclusion is that for the best artificial slip surface with respect to the load support performance, the slip zone must be employed on the leading edge of the parallel sliding contact.

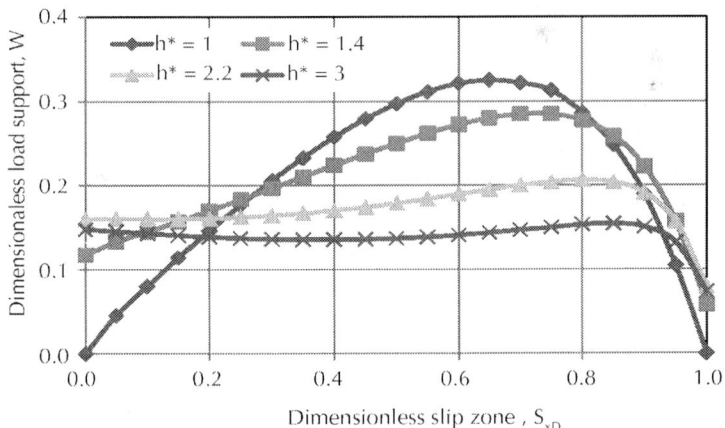

Figure 6: Effect of the dimensionless slip zone of the artificial slip surface S_{xD} on the dimensionless load support W for several wedge ratios h^*.

Figure 7 shows the normalized representation of lubrication film pressure distributions as a function of wedge ratio which are predicted by the modified Reynolds equation (Eq. (1)). For slip configuration, the optimal slip zone S_{xD} of 0.65, is employed. It can be shown that compared with the no-slip boundary condition, the artificial slip surface yields a positive fluid pressure. However, the performance improvement obtained through artificial slip surface is rather mild for the high wedge ratio. Obviously, the lower the base geometry wedge ratio (thus leads to the parallel sliding contact), the larger improvement that boundary slip can induce. In other words, using the artificial slip surface considered here, the maximum pressure occurs not at a convergent wedge as predicted by the classical Reynolds assumption (h^*_{opt} = 2.2 [28]), but at a parallel surface. The predicted maximum pressure for a parallel gap is over three times as large as the maximum pressure obtained from a no-slip contact when the wedge ratio h^* = 2.2. From this perspective, the well-chosen artificial slip surface pattern can be considered as a potential application which is possible for lubricated-MEMS based devices with respect to the load support. In this way, liquid lubricants on the modified opposing surfaces (i.e. artificial slip surface) can prevent the lubrication to break down during device operation to the point where they no longer give proper lubrication.

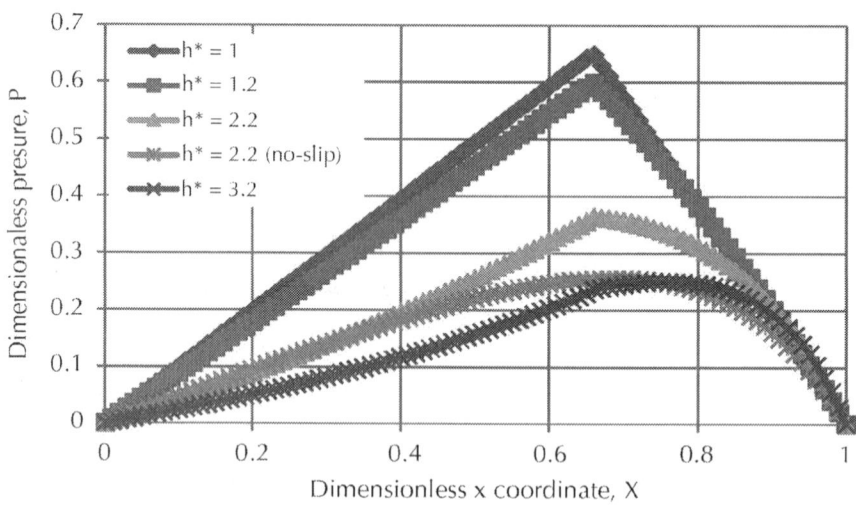

Figure 7: Lubrication film pressure distributions for several values of wedge ratio. The slip profiles are calculated for optimized dimensionless slip zone S_{xD} = 0.65 and dimensionless slip length A = 20.

Figure 8 shows the dimensionless surface friction force F at the bottom surface. It can be seen that the friction of the artificial slip surface becomes smaller than that of a traditional no-slip lubricated contact especially when S_{xD} is larger than about 0.6. This agrees with the numerical analysis of Wu et al. [19] even though the slip model and numerical method used are different. One can remark that when there is no wedge effect (i.e. $h^* = 1$) and $S_{xD} = 0.65$, the artificial slip surface gives the minimum dimensionless friction force of 0.65, while the no-slip lubricated contact gives its dimensionless minimum friction of 0.77 at $h^*_{optimal} = 2.2$, see Table 1. It means that the optimized artificial slip surface of lubricated-MEMS can produce a lower friction force than a no-slip contact.

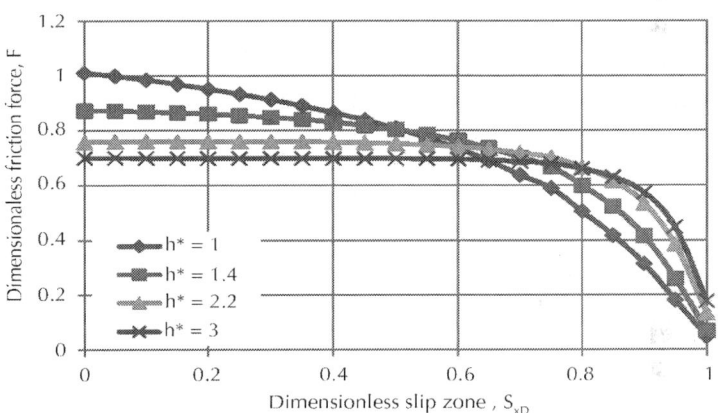

Figure 8: Effect of the dimensionless slip zone of the artificial slip surface S_{xD} on the dimensionless friction force F at bottom surface for several values of wedge ratio h^*. The slip profiles are evaluated for dimensionless slip length A = 20.

Effect of Dimensionless Slip Length on Lubrication Performance

The hydrophobicity of a solid surface, as discussed in the previous section, is usually expressed in terms of a slip length, which quantifies the extent to which the fluid elements near the surface are affected by the surface energy and the surface geometry. The surface energy is an intrinsic property of a material that can be controlled by chemical treatment, such as etching approach and/or coat-on/cast approach. The surface roughness of a hydrophobic solid material can be tuned in order to increment its hydrophobicity and obtain a super-hydrophobic solid surface [33, 34]. In this section, from the numerical point of view, the effect of slip length on the lubrication behavior is studied. The dimensionless slip length is varied from 3 to 300.

Figure 9 shows the effect of slip zone of the artificial slip surface for several dimensionless slip length values on the dimensionless load support. As indicated in Figure 9, the increase in the slip length leads to an increase in the predicted load support. Generally, the larger the slip length at the optimized slip zone of the artificial slip zone, the higher the load support. However, when dimensionless slip length is larger than 30, the dimensionless load support is not affected significantly with the increase in the dimensionless slip length. So, the increase of the load support is not infinitely large. It can be deduced that there is no fluid load support for a lubricated-MEMS when it contains no-slip condition (A = 0). It indicates that the absence of the wedge effect on the pressure generation at parallel lubricated sliding contact has been counterbalanced by the influence of the aritical slip surface application. Again, this condition is very advantageous in engineering a lubricated-MEMS which demonstrates parallel gaps.

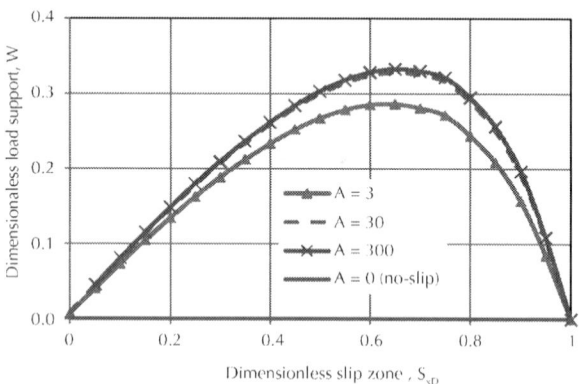

Figure 9: Effect of the dimensionless slip zone of the artificial slip surface S_{xD} on the dimensionless load support W for several values of dimensionless slip length A. All profiles are calculated for parallel sliding surfaces ($h^* = 1$).

In Figure 10 the effect of slip zone of the artificial slip surface for several dimensionless slip length values on the dimensionless friction force is presented. As is well-known, the ability to control and manipulate friction force during sliding is extremely important key to prolong a life-time of lubricated-MEMS. Better understanding of the friction force phenomena at micro-scales is needed to provide designers and engineers the required tools and capabilities to control friction force and predict failure of lubrication in MEMS.

As can be seen in Figure 10 that the artificial slip surface leads to a reduction of the friction force for all dimensionless slip length. The friction force decreases with increasing the slip zone S_x. It can be seen that when the slip zone covers over the whole surface, i.e. homogeneous slip surface, the friction forces has a minimum value, especially for a high slip length. If the reduction in friction force is of only particular interest, the homogeneous slip surface ($S_{xD} = 1$) is very beneficial. But if the performance is also related to the load support, homogeneous slip is not recommended because when $S_{xD} = 1$, the predicted load support is very small for all wedge ratios, see Fig. 6. With respect to the influence of the dimensionless slip length, opposite to the hydrodynamic load support, the dimensionless friction force becomes smaller for higher dimensionless slip length. Therefore, the optimized artificial slip surface is a very promising way to increase the hydrodynamic performance and the stability of the lubricated MEMS system because it gives an advanced load support in combination with a reduced friction force.

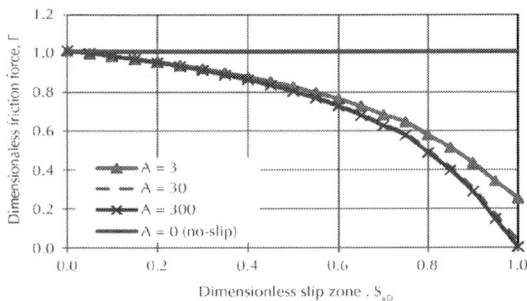

Figure 10: Effect of the dimensionless slip zone of the artificial slip surface S_{xD} on the dimensionless friction force F for several values of dimensionless slip length A. All profiles are calculated for parallel sliding surfaces ($h^* = 1$).

The combined effect of slip zone parameter on load support and friction force can be better analyzed using the dimensionless friction coefficient. In the present study the dimensionless friction coefficient mis defined as the ratio of the dimensionless friction coefficient F to the dimensionless load support W. Figure 11 shows the variation of the dimensionless friction coefficient m as a function of the dimensionless slip zone S_{xD} for various slip lengths A. It appears that when the dimensionless slip zone is smaller than 0.2, the friction coefficient increases significantly. After S_{xD} = 0.2, increasing the slip zone will be less significant to the reduction of the friction coefficient. It can also be deduced from Figure 11 that dimensionless friction coefficient m will decrease with the increase of the dimensionless slip length A, especially for small S_{xD}. It implies that the minimum dimensionless friction coefficient is accord with the highest dimensionless load support (when S_{xD} = 0.65).

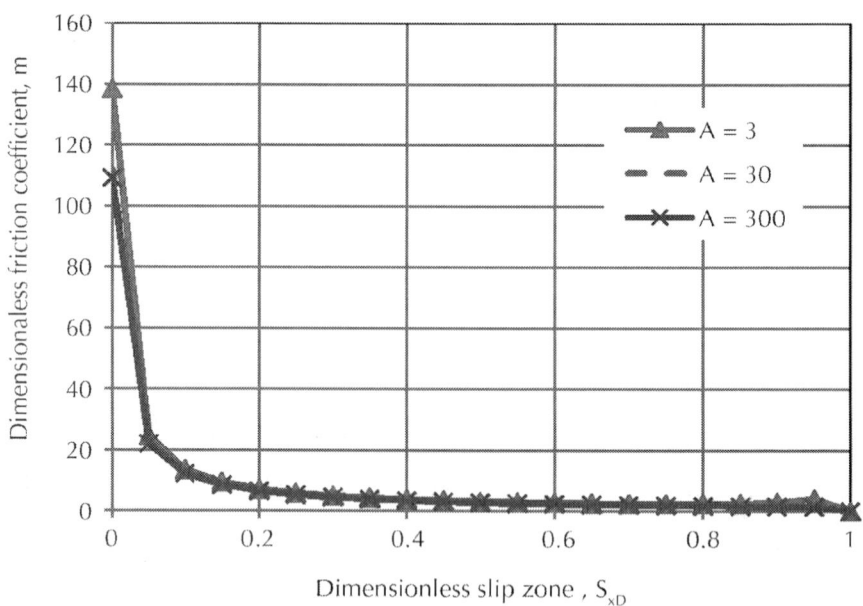

Figure 11: Effect of the dimensionless slip zone of the artificial slip surface S_{xD} on the dimensionless friction coefficient m for several values of dimensionless slip length A. All profiles are calculated for parallel sliding surfaces (h^* = 1).

CONCLUDING REMARKS

Numerical results show that the hydrodynamics of a lubrication film confined between a moving no-slip surface and a stationary with an artificial slip surface differ significantly from that of a film confined between two no-slip surfaces. It is found that a homogeneous slip boundary on one surface produces a lower hydrodynamic pressure in a lubricated sliding contact at various conditions (slope incline, and slip length), resulting in a reduced load support which reduces the positive effect of slip on friction. However, if the surface is designed with an optimal artificial slip pattern (the slip zone is applied on 0.65 of contact length), even when there is no wedge effect, the load support has a maximum value. In addition, the friction force can decrease significantly. Therefore, it is very beneficial to make one of the contacting surfaces in lubricated-MEMS with an artificial slip surface for achieving ideal lubrication performance, i.e. reduced friction coefficient and increased load support.

REFERENCES

1.   Houston MR, Howe RT, Maboudian R. Effect of Hydrogen Termination on The Work of Adhesion Between Rough Polycrystalline Silicon Surfaces. Journal of Applied Physics 1997;81(8): 3474–3483.

2.   Maboudian R, Ashurst WR, Carraro C. Self-Assembled Monolayers as Anti-Stiction Coatings For MEMS: Characteristics and Recent Developments. Sensors and Actuators A: Physical 2000;82(1–3): 219–223.

3.   Tagawa M, Ikemura M, Nakayama Y, Ohmae N. Effect of Water Adsorption on Microtribological Properties of Hydrogenated Diamond-Like Carbon Films. Tribology Letters 2004;17(3): 575–580.

4.   Smallwood SA, Eapen KC, Patton ST, Zabinski JS. Performance Results of MEMS Coated with a Conformal DLC. Wear 2006;260: 1179–1189.

5.   van Spengen WM, Puers R, De wolf I. On The Physics of Stiction and Its Impact on The Reliability of Microstructures. Journal of Adhesion Science and Technology 2003;17(4): 563–582.

6.   Pit R, Hervet H, Leger L. Direct Experimental Evidence of Slip in Hexadecane: Solid Interfaces. Physical Review Letters 2000;85: 980–983.

7.   Craig VSJ, Neto C, Williams DRM. Shear-Dependent Boundary Slip in an Aqueous Newtonian Liquid. Physical Review Letters 2001;87(054504).

8.   Bonaccurso E, Kappl M, Butt HJ. Hydrodynamic Force Measurements: Boundary Slip of Hydrophilic Surfaces and Electrokinetic Effects. Physical Review Letters 2002;88(076103).

9.   Bonaccurso E, Butt HJ, Craig VSJ. Surface Roughness and Hydrodynamic Boundary Slip of A Newtonian Fluid in A Completely Wetting System. Physical Review Letters 2003;90(144501).

10.   Zhu YX, Granick S. Rate-Dependent Slip of Newtonian Liquid at Smooth Surfaces. Physical Review Letters 2001;87(096105).

11.   Priezjev NV, Darhuber AA, Troian SM. Slip Behaviour in Liquid Films on Surfaces of Patterned Wettability: Comparison Between Continuum and Molecular Dynamics Simulations. Physical Review E 2005;71(041608).

12.   Cottin-Bizonne C, Barentin C, Charlaix E, Bocquet L, Barrat JL. Dynamics of Simple Liquids at Heterogeneous Surfaces: Molecular Dynamics Simulations and Hydrodynamic Description. European Physical Journal E 2004;15: 427– 438.

13.   Harting J, Kunert C, Herrmann HJ. Lattice Boltzmann Simulations of Apparent Slip in Hydrophobic Channels. Europhysic Letters 2006; 75: 328–334.

14.   Li BM, Kwok DY. Discrete Boltzmann Equation for Microfluidics. Physical Review Letters 2003;90(124502).

15.   Spikes HA. The Half-Wetted Bearing. Part 1: Extended Reynolds Equation. Proceedings of the Institution of Mechanical Engineers, Part J: Journal of Engineering Tribology 2003; 217: 1–14.

16.   Spikes HA. The Half-Wetted Bearing. Part 2: Potential Application in Low Load Contacts. Proceedings of the Institution of Mechanical

Engineers, Part J: Journal of Engineering Tribology 2003;217: 15–26.

17. Salant RF, Fortier AE. Numerical Analysis of A Slider Bearing with A Heterogeneous Slip/No-Slip Surface. Tribology Transactions 2004; 47: 328–334.

18. Fortier AE, Salant RF. Numerical Analysis of A Journal Bearing with A Heterogeneous Slip/No-Slip Surface. ASME Journal of Tribology 2005;127: 820–825.

19. Wu CW, Ma GJ, Zhou P. Low Friction and High Load Support Capacity of Slider Bearing with A Mixed Slip Surface. ASME Journal of Tribology 2006;128: 904–907.

20. Choo JH, Glovnea RP, Forrest AK, Spikes HA. A Low Friction Bearing Based on Liquid Slip at The Wall. ASME Journal of Tribology 2007;129: 611–620.

21. Tauviqirrahman M, Ismail R, Jamari, Schipper DJ. Effect of Boundary Slip on The Load Support in A Lubricated Sliding Contact. AIP Conference Proceedings 2011;1415: 51-54. doi:10.1063/1.3667218.

22. Aurelian F, Patrick M, Mohamed H. Wall Slip Effects in (Elasto) Hydrodynamic Journal Bearing. Tribology International 2011;44: 868–877.

23. Tauviqirrahman M, Ismail R, Jamari, Schipper DJ. Wall Slip Effects in a Lubricated MEMS. International Journal of Energy Machinery 2011;4(1): 13–22.

24. Navier CLMH. Mémoire Sur Les Lois Du Mouvement Des Fluides. Mémoires de l'Académie Royale des Sciences de l'Institut de France 1823:6: 389–440.

25. Israelachvili J. Intermolecular and Surface Force, vol. 1, second ed. Academic Press, London; 1995.

26. Reynolds O. On The Theory of Lubrication and Its Application to Mr. Beauchamp Tower's Experiments, Including An Experimental Determination of The Viscosity of Olive Oil. Philosophical Transactions of the Royal Society of London, Part I 1886;177: 157–234.

27. Patankar SV. Numerical Heat Transfer and Fluid Flow. Taylor & Francis, Levittown 1980:30–58.

28. Cameron A. The Principles of Lubrication. Longman Green and Co., ltd; 1966.

29. Kompvopoulos K. Surface Engineering and Microtribology for Microelectromechanical System. Wear 1996;200: 305–327.

30. Tretheway DC, Meinhart, CD. Apparent Fluid Slip at Hydrophobic Microchannel Walls. Physics of Fluids 2002;14: L9-12.

31. Watanabe K, Yanuar, Udagawa H. Drag Reduction of Newtonian Fluid in a Circular Pipe with a Highly Water-Repellant Wall. Journal of Fluid Mechanics 1999;381: 225–238.

32. Ou J, Perot B, Rothstein JP. Laminar Drag Reduction in Microchannels using Ultrahydrophobic Surfaces. Physics of Fluids 2004;16: 4635.

33. Patankar NA. On the Modelling of Hydrophobic Contact Angles on Rough Surfaces. Langmuir 2003;19: 1249–1253.

34. Lafuma A, Quere D. Superhydrophobic States. Nature Materials 2003;2: 457–460.

Chapter 5

Friction and Wear of a Grease Lubricated Contact — An Energetic Approach

Erik Kuhn[1]

[1]Department of Mechanical Engineering and Production, Hamburg University of Applied Sciences, Germany.

INTRODUCTION

Lubricating greases are colloid disperse systems with visco-elastical properties. These special lubricants have a wide range of application. More than 90% of the ball bearings are grease lubricated but also gears and journal bearings are application examples for greases.

The composition of greases consists of a base oil and a thickener (of course some additives can be found in commercial lubricants). The thickener forms a network which leads to a complex rheological and tribological behaviour.

Tribological contacts lubricated by greases are often working under mixed friction conditions. The situation inside the grease film determines the content of fluid friction. The situation inside the tribological gap influences the content of solid friction. A special phenomenon is the structural degradation due to the effects of friction. It leads to a dependence on time of the grease behaviour.

Aims of this chapter are the description of the liquid friction process inside the grease film and of the procedure of structural degradation. Both phenomena have a strong influence on the friction and wear behaviour of the whole tribo-system.

GENERAL DEFINITIONS AND TERMINOLOGY

The presented work is based on a conception of friction and wear that investigates the process from an energy point of view. Some ideas differ from well-known definitions so that they need to be introduced here for better understanding.

A *tribological contact* describes the geometrical situation of rubbing bodies with interactions in the sense of the tribological process.

Friction is (only) an energy expenditure.

Wear is a production of irreversibility due to affected friction energy. It covers all elements of a tribo-system [1].

This chapter distinguishes between wear volume and volume of the removed material.

Wear volume is presented by the material area where irreversible friction effects lead to an excess of a critical energetic level. (n contrast to the removed material volume (loss of material)).

Mixed friction describes an energy expenditure in several states of friction that exists simultaneously inside the same tribo-system (see also [2]). A direct contact of the solid bodies is not necessary for mixed friction of a lubricated couple (mix of solid and liquid friction).

Lubricating greases are colloid disperse systems with visco-elastical properties.

The grease structure is characterised by the geometry and the distribution of the thickener, the interactions between the thickener and the tribo-system and the ability of storing energy.

ENERGY BALANCE FOR A GREASE LUBRICATED CONTACT

Grease lubricated contacts are often working in mixed friction. That means liquid friction (grease film) and solid friction (asperity deformation) have to be considered. An idea of the situation inside the lubricated contact is presented in the next figures.

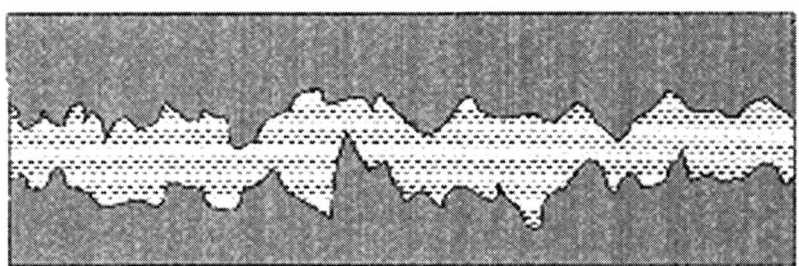

Figure 1: Grease lubricated contact with two rough surfaces.

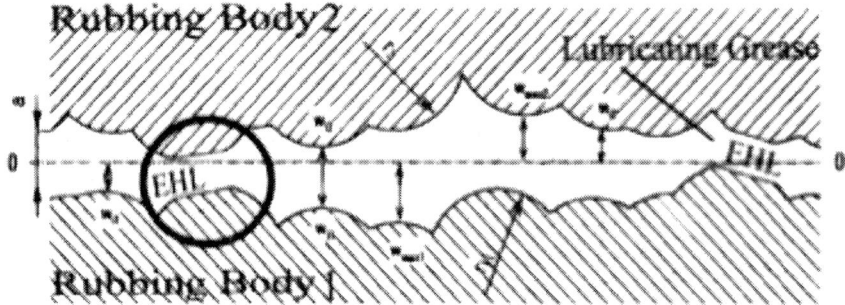

Figure 2: Modelled mixed friction contact of a grease lubricated gap.

The surface profile is modelled with a spherical shape of the asperities and the gap configuration can be illustrated with Fig.2. The situation of a single contact in consideration of the simple idea of lubricant layers inside the grease film is presented in Fig.3.

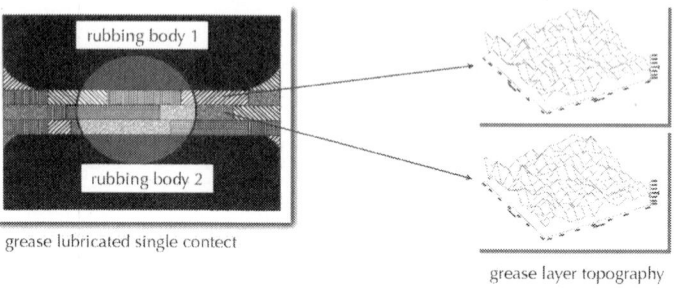

grease lubricated single contact

grease layer topography

Figure 3: Single contact and modelled contact situation inside the grease film.

The grease topography comes from IR-microscopy and presents the density distribution. In Fig.3 the contact of two density areas inside the greases film is highlighted.

Investigations of the surface profile and of the density distribution of the grease lead to a discrete contact model. Random variables are asperity height $_{1,2}$, radius of the modelled asperity shape $_{1,2}$ and the observed density $_{1,2}$. All random parameters can be described with the Gaussian distribution.

For the contact probability F illustrated in Fig.3 (left) can be obtained [3]

$$F(z,u,\rho) = \int_0^z \int_0^{z-\xi_2} f(\xi_1)f(\xi_2)d\xi_1 d\xi_2 \cdot \int_{uk1}^{ug1} f(\rho_1)d\rho_1 \cdot \int_{uk2}^{ug2} f(\rho_2) \cdot \int_{pk1}^{pg2} f(\delta_1)d\delta_1 \cdot \int_{pk2}^{pg2} f(\delta_2)d\delta_2 \quad (1)$$

An energy balance is created by investigation of a single grease lubricated contact. The expended energy is formed as a sum of different contents

$$W_{friction} = \sum_i W_{i=1..n} \quad (2)$$

The assumption of mixed friction leads to the consideration of two main contents of the friction energy.

$$W_{mixed-friction} = W_{solid} + W_{fluid} \qquad (3)$$

A proposal is made with

$$W_{solid} = e_{elast/plast} \cdot V_{elast/plast} \cdot n_{elast/plast} \qquad (4)$$

$$W_{liquid} = e_{rheol,2,3} \cdot V_{rheol,2,3} \cdot n_{rheol,2,3} + e_{solidif} \cdot n_{solidif} \cdot V_{solidif} + e_{tensile} \cdot V_{tensile} \cdot n_{tensile} + e_{wave} \cdot V_w \\ _{ave} \cdot n_{wave} \qquad (5)$$

In Eq. (4) and (5) the notation *energy density* for the selected mechanism multiplied by *stressed volume* and by the number of contacts with the same mechanism is used. For a single contact n=1 holds. The solid friction from Eq. (4) considers the deformation (elastically/plastically) of the stressed asperities. Information about this is given in [4].

The summands in Eq.(5) describe the shearing process in different gap situations (index *rheo*), solidification effects in the center of the observed gap (index *solidif*), tensile stress in the outlet of the contact (index *tensile*) and the formation of a friction wave (index *wave*).

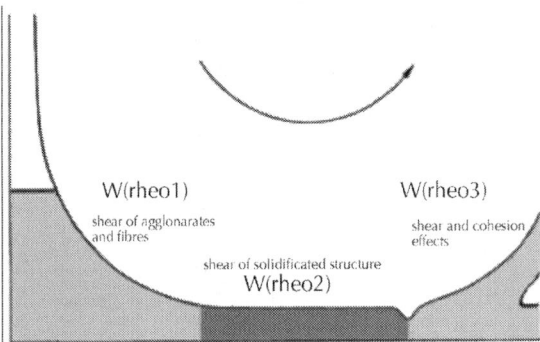

Figure 4: Different stress situations for the grease inside the contact geometry (erheo1, 2, 3).

LIQUID FRICTION INSIDE THE GREASE FILM

Stress Situation inside the Gap Geometry

As mentioned before a micro single contact of a grease lubricated couple is observed and the stress situation is analysed. The idea of the process model developed here is illustrated in Fig.5.

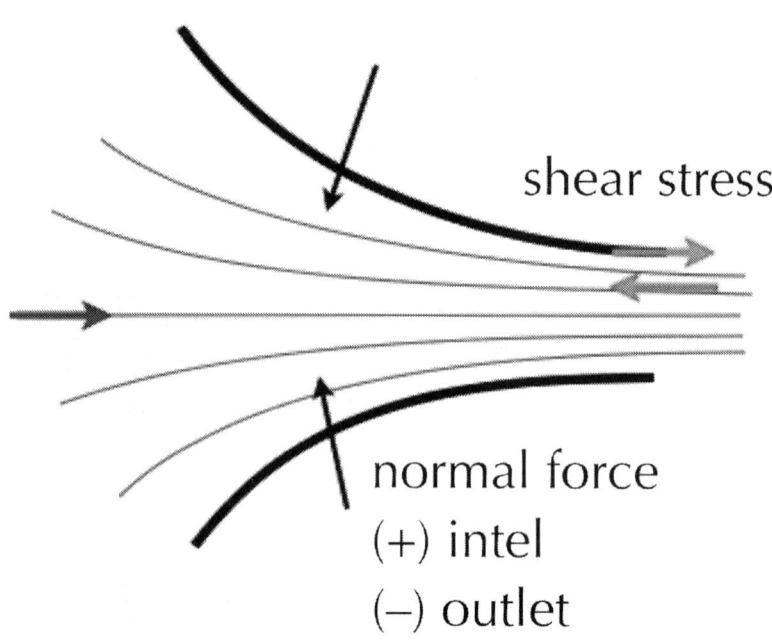

shear stress

normal force
(+) intel
(–) outlet

Figure 5: Grease lubricated gap between two asperities (micro single contact). Shear stress is the most important stress mechanism of the grease. Normal force (+) leads to solidification effects. Normal force (-) leads to tensile stress.

In addition the possibility of a friction wave inside the grease film is observed. The idea is clarified in the Fig.6.

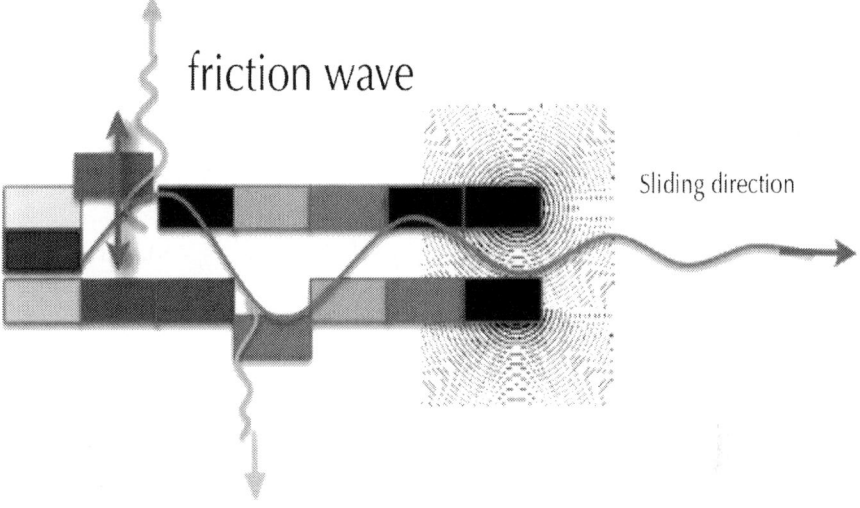

Figure 6: Schematic representation of a friction wave caused by the contact of volume elements of the grease film with different properties.

Empirical Proposals to Quantify the Friction Energy

The experimental work is focused on the quantification of the energy expenditure during the shear process of a lubricant. This requires an experimental procedure that simulates the liquid friction inside the grease film.

A proposal is made by the use of a rheometer [5], [6]. Shearing a grease sample in a rotating or oscillating test can be interpreted as a fluid friction experiment. All reaction measured by the rheometer is caused by the grease behaviour. The plate and cone in a rheometer configuration does not lead to the state of mixed friction. Although the test conditions are fare from real contact situation rheometer tests are helpful for fundamental investigations of fluid friction.

A picture of a cone-plate configuration is shown in Fig.7.

Figure 7: Evolution of the observed shear stress (left) during a rheometer experiment with a cone-plate system (right) in rotational modus.

The energy per volume that is necessary to shear the grease during the experiment can be obtained from a shear test (Fig.7). The experimental conditions are constant shear rate and constant test temperature. To compare different grease samples the test time for each experiment has to be held constant.

An empirical proposal [7] is made with

$$e_{rheo} = \dot{\gamma} const \cdot \int_0^t \tau(\varsigma) d\varsigma \qquad (6)$$

with e_{rheo} the energy density for the shear process [J/m³], t the test time [s], $\dot{\gamma}$ the shear rate [1/s] and the current time [s].

A more interesting experimental procedure is an oscillating rheometer test [Fig.8]. The grease behaviour can be observed in a wide range of the oscillating amplitude. For the investigation of fluid friction inside the grease film experiments within the linear visco-elastical range can be analysed.

To quantify the friction behaviour the following proposal [3] can be used

$$e_{rheo-elast} = \frac{G' \cdot \gamma elast^2}{\cos\delta} \qquad (7)$$

G' is the storage modulus [Pa], is the deformation [-] and is the phase different angle. Eq., (7) describes the energy per volume [J/m³] to deform the sample elastically and is related to the thickener.

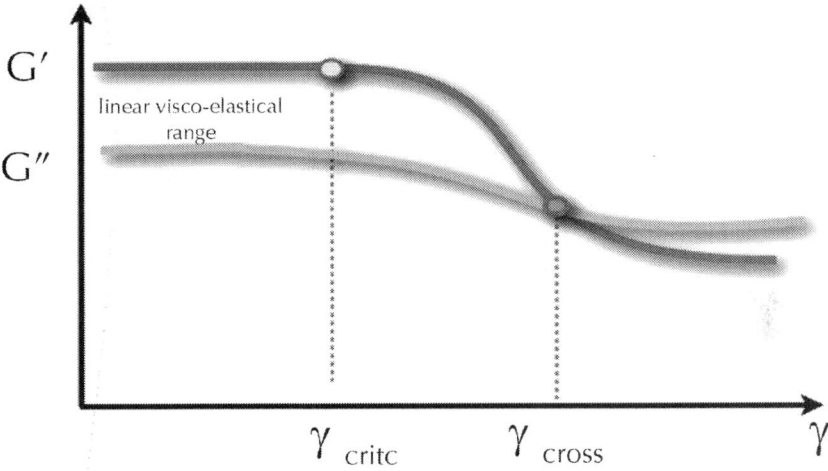

Figure 8: Typical evolution of the storage modulus and the loss modulus during an amplitude sweep (oscillating measurement).

Results from Experimental Work

The energy expenditure expressed by the rheological energy density $e_{rheo-elast}$ presents the liquid friction behaviour of the investigated grease samples. Equation (7) observes only the shear mechanism inside the tribological gap. To compare the greases the same deformation (oscillating amplitude) has to be used. Some greases with the same thickener type (Li-soap) and the same base oil (mineral oil) were observed by a variation of soap content and test temperature.

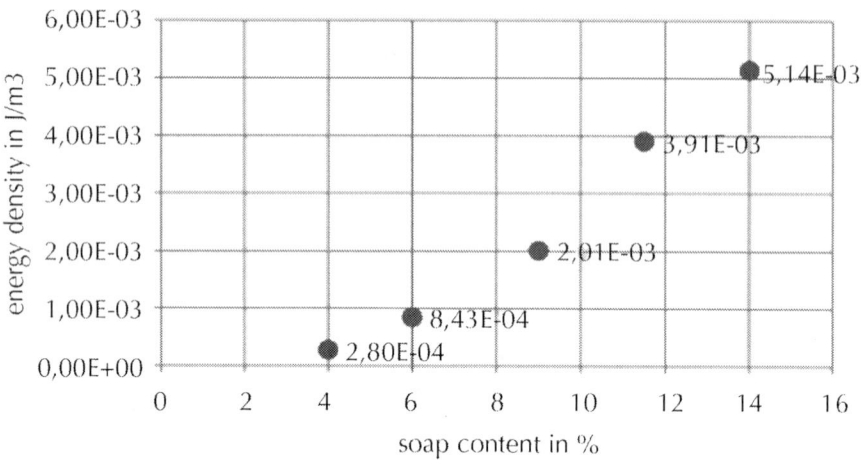

Figure 9: Energy densities from the linear visco-elastical range =20°C for different soap.

An increase of the soap content leads to an increase of the liquid friction. Experiments with a temperature =50°C and the same grease samples deliver the results in Fig.10.

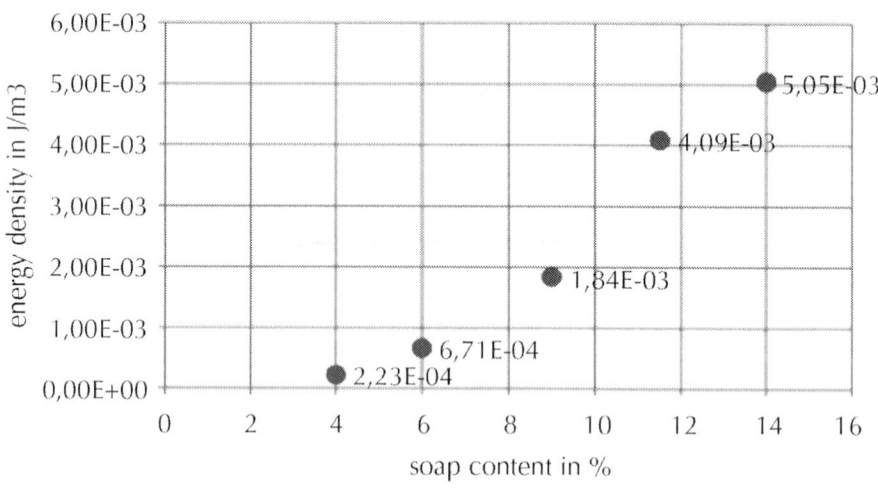

Figure 10: Energy densities for a test temperature =50°C (same conditions as Fig.9)

Compared with the test temperature of =20°C lower values of the energy densities are obtained. This behaviour is in accordance with the experience that fluid friction is decreasing with an increasing temperature. A more or less linear correlation of energy density and soap content can be observed.

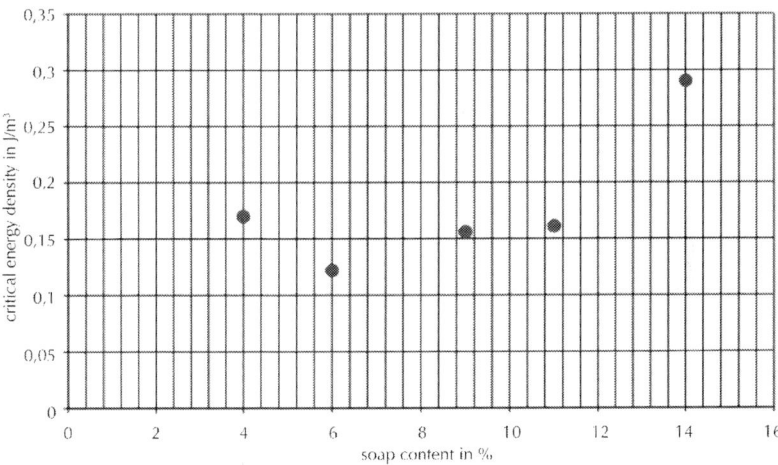

Figure 11: Critical energy level vs. soap content.

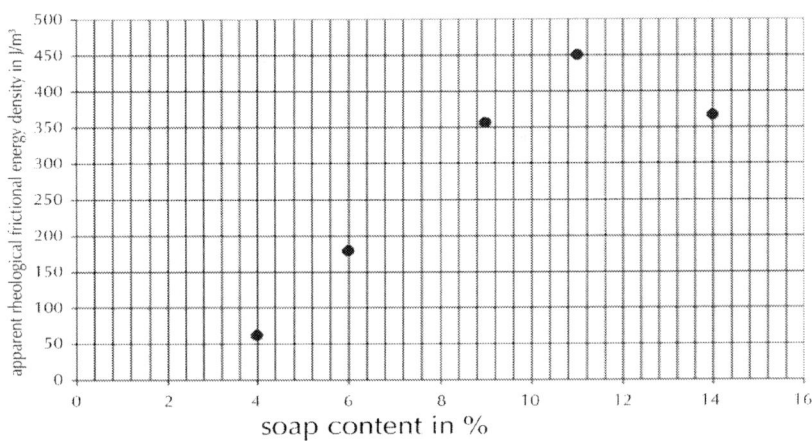

Figure 12: Energy densities for the crossing point vs. soap content

IRREVERSIBLE EFFECTS DUE TO FRICTION

IDEA OF THE STRUCTURAL DEGRADATION OF LUBRICATING GREASES

A typical curve obtained from rheometer experiments for constant shear rate and temperature (rotational mode) shows a strong dependence on time. The drop of shear stress versus stress time is an indirect expression of the structural degradation and well known from many papers [8], [9], [10],.

To illustrate the friction effects AFM-investigations made by [11] are presented below (Fig.13). The change of thickener structure caused by the liquid friction is evident. The geometry and distribution of the thickener is completely different to the initial situation and it can be assumed that the new grease structure shows a different tribological behaviour.

Volume elements inside the grease film are modelled to observe the tribological process. Because of the thickener distribution these volume elements have different properties as elasticity, density, level of accumulated energy, level of critical energy et. The consequence of the property distribution is a different tribological behaviour of the observed volume elements forming a lubricant layer. The contact situation of two assumed grease layer composed of different volume elements is presented in Fig. 14.

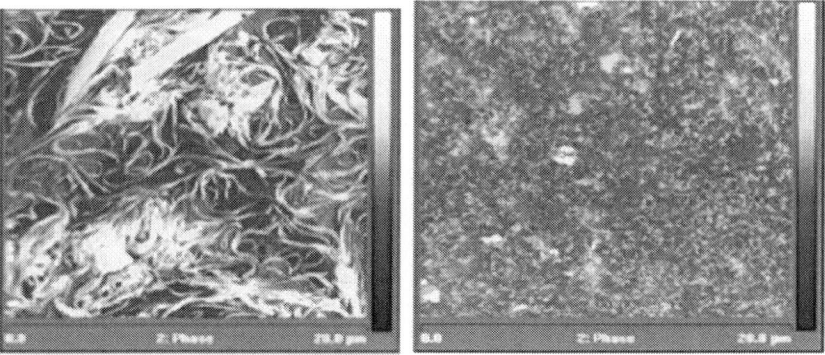

Figure 13: Left- fresh grease sample, right- grease stressed in a rheometer [11]

Liquid friction between two modelled grease layer. Different volume elements have different rheological and tribological properties.

An energy stress is applied to the control volume if liquid friction take place (Fig.15). Due to this energy stress an energy accumulation process, a dissipation process and a transition process (overstep of a critical energy level) starts.

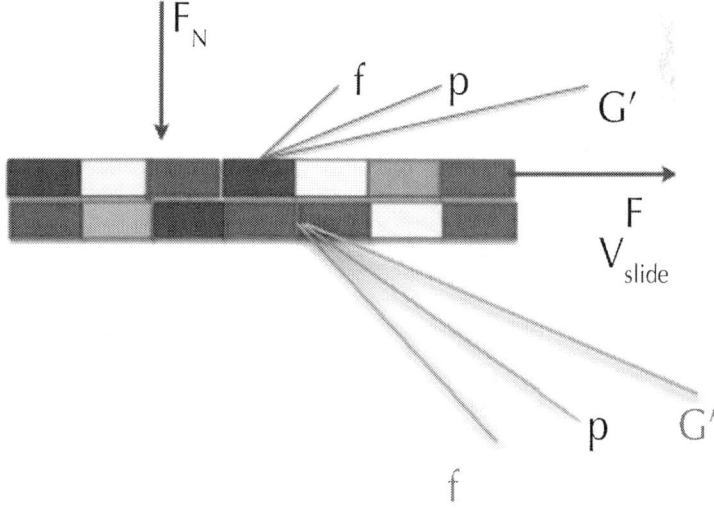

Figure 14: Liquid friction between two modelled grease layer Different volume elements have different rheological and tribological properties.

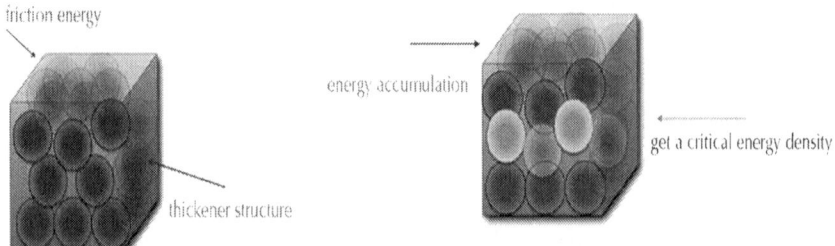

Figure 15: Modelled volume element inside the grease film. Left – unstressed, right – stressed by liquid frictio

The degradation of the grease structure begins with a transition process within a volume element. Overstepping a critical energy density initiates an irreversible change of the structure.

A contact model is developed to describe the energetic situation of critical exceedance.

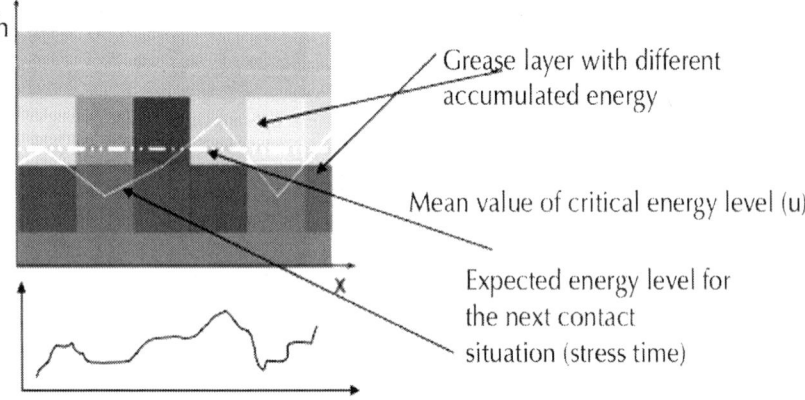

Figure 16: New model to describe the energetic situation for the transition process inside the grease film [1]

To quantify the number of exceedance of critical energy level a stationary Gaussian process is used. Information about the conditions and definitions can obtained in [1] and [12].

$$E[N_0(u)] = \lim_{\varepsilon \to 0} \frac{1}{\varepsilon \cdot L} \cdot \int_0^L E[A(x,\varepsilon) \cdot h'(x)]dx \qquad (8)$$

$$E[N_0(u)] = \frac{1}{2\pi} \cdot \sqrt{\frac{m_2}{m_0} \cdot e^{\frac{u^2}{2m_0}}} \qquad (9)$$

Initial situation is described with Eq. (8) and the expectation of the number of overstepping

N_0 can be determined with Eq. (9). This proposal uses spectral moments m_2; m_0. An example [1] was quantified by using density distribution from IR-microscopy to determine the parameter m_2;m_0. A mean value of the critical energy level was assumed.

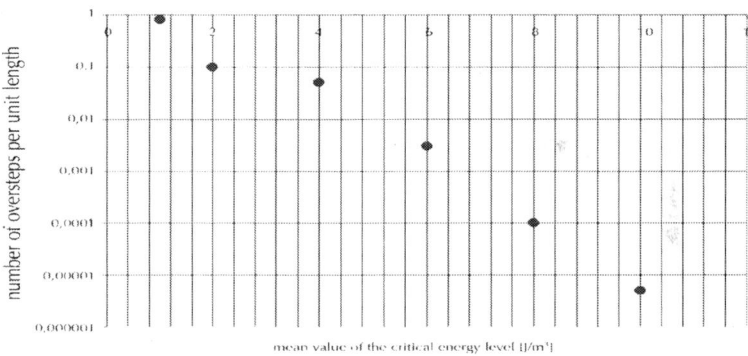

Figure 17: Influence of the critical energy level of the number of oversteps [1]

Thermodynamic Investiagtions

General Aspects

Friction process within a tribo-system is an irreversible process. It means that input of friction energy leads to irreversible effects. This approach interpreted friction and wear process as an cause-effect-chain.

Different authors tried to find relations between the system behaviour expressed by mass loss (wear) and entropy flow/production [15]-[19]. *Abdel-Aal* [18] expressed the conjecture that relation between frictional heat generation and heat dissipation is related to wear transition. This author pointed out [19] that there exists a one to one relationship between entropy generation and mass loss. *Doelling et al.* [17] give the argumentation that there exists a strong correlation between components wear and entropy flow.

Ling et al. [13] give an experimental description of a correlation between wear and entropy flow in lubricated sliding systems.

„Sliding wear is an irreversible degradation of surfaces induced by friction. On a microscopic scale irreversible physical interactions between the sliding surfaces – including plastic deformation of asperities, fracture, delamination, abrasive plowing, and corrosive wear, among others – creates friction resistance forces, dissipates power, and generates irreversible entropy. Since the physical interactions responsible for friction and wear monotonically produce entropy, entropy becomes a time base for wear" [13].

"The entropy production, in fact, enable one to bridge the atomic scale phenomena with the macro scale response" [14] in [13]. "In addition friction and wear, from the vantage point of thermodynamics irreversible transform mechanical energy into other forms through dissipative processes. Therefore, entropy production is believed to be a propitious measure for a systematic study of wear and friction" [15].

All these investigations pointed out that the application of irreversible thermodynamics is a promising tool to analyse the tribological processes.

Entropy and Structural Degradation

The aim of this chapter is a description of relation between energetic situation of the tribo-system and the degradation of grease structure. Tribo-sytems as solid surface 1, solid surface 2 and lubricant are investigated but also subsystems as lubricant layer 1 against lubricant layer 2.

Source of irreversible processes are thermodynamic forces X_i (i=1, 2...) (gradient of temperature, gradient of concentration...). These forces X_i evoke corresponding flows I_i (i=1,2,...) (heat flow, diffusion flow...) The generalisation of classical thermodynamic to describe irreversible effects leads to an investigation of local equilibrium. That means there exist macroscopic small system areas that are provided in equilibrium while the whole system is out of equilibrium. Two principles of thermodynamic are used: the linear dependence of flows and thermodynamic forces, and the *Onsager-reciprocity* [20].

For the entropy generation can be written

$$\frac{dS}{dt} = \sum_i I_i \cdot X_i$$

The variation of entropy is influenced by two terms

$$d_S = dS_{out} + dS_{in} \quad (10)$$

Heat transfer across the boundary of the modelled system (subsystem) delivers the entropy dS_{out}. Entropy related to mechanisms taking place inside the system is described with dS_{in} (entropy production). The intention is to relate the entropy production inside the system with different irreversible effects caused by the friction process within the grease layer. It can be written

$$\rho \cdot \frac{dS}{dt} + div\sigma = \Theta \qquad (11)$$

with - entropy flow density ($= \acute{A}/T$; $\acute{A}$ =heat flow density) and - local entropy increase per unit time ($=-(\acute{A},gradT)/T^2$) [20]

Written in differential form and related to time

$$\frac{dS}{dt} = \frac{dS_{in}}{dt} - \frac{dQ_{1-2}}{dt} + s_e \cdot \frac{dm_e}{dt} - s_a \cdot \frac{dm_a}{dt} \qquad (12)$$

with $s_e \cdot \dfrac{dm_e}{dt}$ - entropy transported into the system with mass transport; $s_a \cdot \dfrac{dm_a}{dt}$ - entropy transported out of the system by mass loss.

S_{in} describes the entropy production inside the system and $(\pm)SQ_{1-2}$ leads to a change of system entropy by heat transfer across the system boundaries.

Eq. (12) describes the entropy balance for an open thermodynamic system (with exchange of mass). As Abdel-Aal [18] pointed out only the entropy source strength, namely entropy created in the system, should be used as a basis for systematic description of the irreversible process (degradation of materials).

The process of structural degradation can be described as a process of energy accumulation, energy dissipation and transition of critical energy levels. Each of these mechanisms delivers a contribution to an entropy balance and will change the production term in Eq.(10) and (12). Entropy production (for the observed volume element) is determined by fluid friction and its effects. The process of solid friction (for a mixed friction contact) delivers only a heat portion into the modelled system. The tribo-subsystem can be illustrated with Fig.18.

The transport processes modeled from the investigated tribo-system are presented in Fig. 19, 20 and21.

As a consequence of the modelled mechanisms for the entropy production can be written as

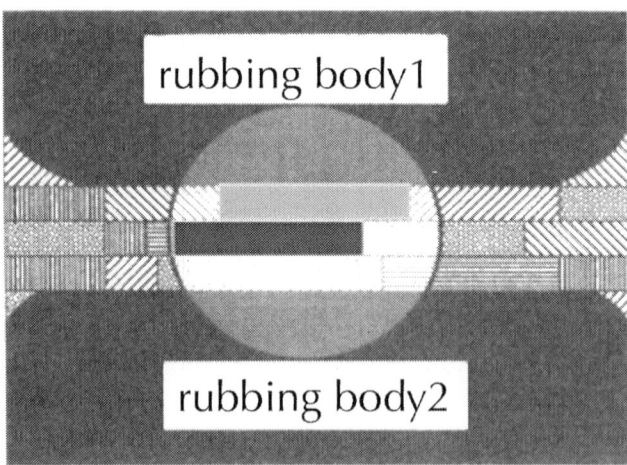

Figure 18: Tribo-subsystem (grease layer against grease layer) inside the general tribo-system.

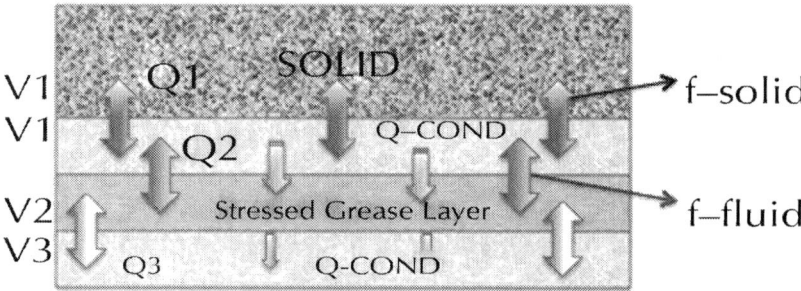

Figure 19: Observed contact situation (section of a complete tribological system). Solid rubbing body and some modelled grease layers. Heat transport: Q1 - heat amount from solid friction transported into the solid material and into the grease layer; Q2 - heat amount from liquid friction flows into layer 1 and 2; Q3 - same as Q2; Q_{COND} presents the heat flow by conduction from layer to layer; V - velocities of the observed layers.

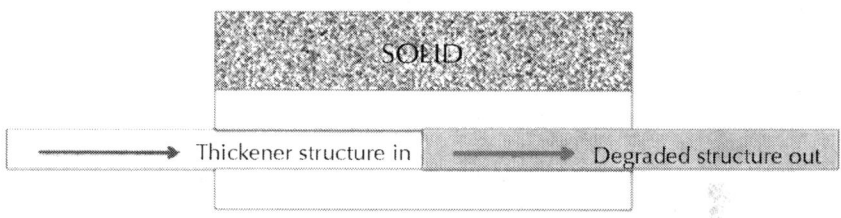

Figure 20: Imagine of tEransport of thickener structure in and out of the observed volume element (mass transport).

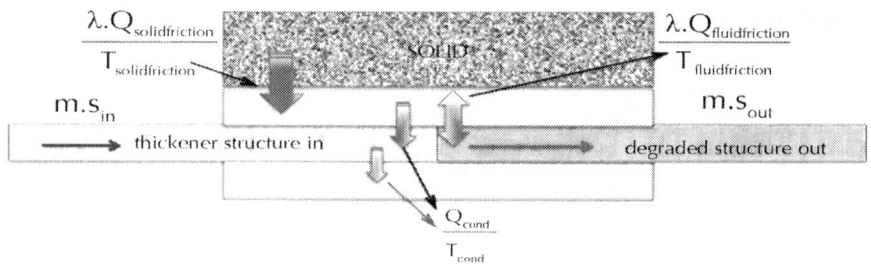

Figure 21: Modelled entropy transport processes inside the observed system (entropy flow) [22]

As a consequence of the modelled mechanisms for the entropy production can be

Written with

$$\frac{dS_{in}}{dt} = \frac{dS_{acc}}{dt} + \frac{dS_{diss}}{dt} + \frac{dS_{trans}}{dt} \tag{13}$$

It means the process of energy accumulation the process of energy dissipation and the transition of an critical energy level produce entropy. Any chemical potential is disregarded in this investigation. We may rewrite Eq. (12)

$$\frac{dS}{dt} = \left(\frac{dS_{acc}}{dt} + \frac{dS_{diss}}{dt} + \frac{dS_{trans}}{dt}\right) - \frac{dQ_{1-2}}{dt} + s_e \cdot \frac{dm_e}{dt} - s_a \cdot \frac{dm_a}{dt} \tag{14}$$

An assumption is made that heat flow gets a (−) that means for a balance that heat leaves the observed volume element.

It can be proposed

$$S_{acc} = \frac{e_{def} \cdot \varsigma_R \cdot V_{acc}}{T_{acc}} \tag{15}$$

with e_{def} the energy density used for deformation process, $_R$ describes the part of friction energy which is accumulated, V_{acc} the accumulation volume and T_{acc} the temperature of accumulation process.

$$S_{trans} = \frac{G' \cdot \gamma_{critic}^2}{\cos\delta} \tag{16}$$

from oscillating rheometer measurements (see Eq.(7)). The temperature situation is assumed as $T_{solid} > T_{layer1} > T_{layer2}$. Furthermore it is assumed that different temperature appears for different mechanism. It means

$$T_{acc} \neq T_{diss} \neq T_{trans}.$$

In general it is conceivable that part of heat generated by solid friction (asperity deformation) enters the first modelled grease layer. Part of this thermal load will be conducted into the next lubricant layer. Liquid friction between the modelled layers leads to a transport of heat into the layers.

An interesting description of the entropy production term for different processes at sliding interfaces comes from [21].

To link the energetic situation with the structural degradation the entropy production term has to be simplified. The friction energy W_f is observed with the temperature T_f. It can be obtained

$$e^*_{Rrheo} = T_f \cdot (\rho_a \cdot s_a) - \frac{T_f}{V_a}(S_e - S_{Q1-2}) \tag{17}$$

An interpretation delivers ($_a \cdot s_a$) as an entropy density leaving the system with the mass exchange. It means an increasing energetic release by entropy flow out of the system with the degraded structure leads to an increasing capacity to withstand stresses expressed by e^*_{Rrheo}.

To use some experimental results some assumption were made. The specific entropy can be determined by

$$s = c \cdot \ln\left(\frac{T_{str}}{T_{unstr}}\right) \tag{18}$$

with index *str* for the stressed layer and *unstr* for the unstressed layer. Test temperature in the rheometer was used for T_{unstr}. An increasing temperature with an increasing soap content during the friction process was assumed because all measurements show a direct correlation between soap content and fluid friction (see Fig. 10). To link the specific entropy leaving the system with the degradation process the parameter e^*_{Rrheo} for the crossing point (rheometer tests) is used.

Fig. 22 presents the tendency of e^*_{Rrheo} vs. s.

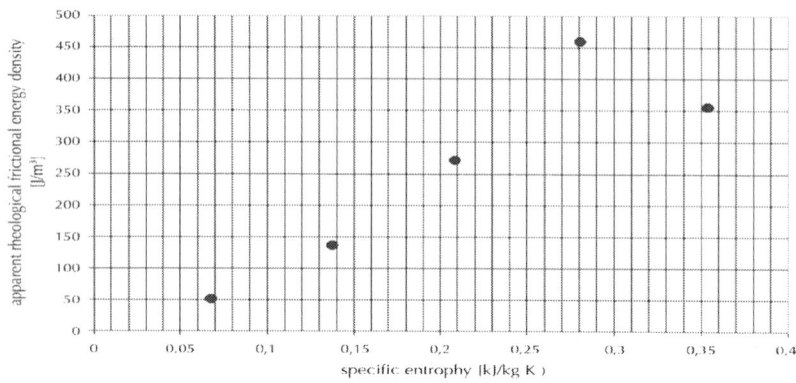

Figure 22: Correlation between e^*_{Rrheo} (the crossing point) and specific entropy.

Fig. 22 presents the assumed correlation between energetic stress and energetic release.

Finally some conclusions were made

$$\frac{dS}{dt} > \dot{S}_{prod} + \dot{S}_{Q1-2} + \dot{m} \cdot s_e \qquad (19)$$

- A significant energetic release by the transport of degraded structure out of the system can be observed.

$$\frac{dS}{dt} = \dot{S}_{prod} + \dot{S}_{Q1-2} + \dot{m} \cdot s_e \qquad (20)$$

- No significant energetic release by the transport of degraded structure out of the system can be observed.

$$\frac{dS}{dt} < \dot{S}_{prod} + \dot{S}_{Q1-2} + \dot{m} \cdot s_e$$

- An additional entropy source happens (for example the mentioned friction wave).

CONCLUSIONS

Some new definitions of general tribological subjects are made. A proposal for an energy balance of a grease lubricated contact is given and empirical proposals for quantification are presented. With the help of rheometer tests the fluid friction was investigated and results are illustrated. The degradation process of the grease structure is described with energetic parameters. An open thermodynamic system is created and described. The influence of the energy flow on the degradation process is presented too.

REFERENCES

1. Kuhn,E.: Tribology of lubricating greases. An energetical approach of the tribological process (in German). expert verlag, 2009

2. Fleischer,G.: Criterions of mixed friction (in German). Tribologie und Schmierungstechnik, Berlin Technik Verlag 1985

3. Kuhn,E.: Investigation of the Structural Degradation of Lubricating Greases due to Tribological Stress. Intern. Colloquium Tribology Esslingen, 2012.

4. Fleischer,G.: Scientific findingy by Tross from a contemprorary point of view. (in German) 3rd Arnold Tross Colloquium, Hamburg 2007

5. Kuhn,E.: Inherent tribo-system response to optimise the process conditions. 8th Arnold Tross Colloquium, Hamburg 2012

6. Kuhn,E.: Tribological stress and structural behaviour of lubricating greases. ECOTRIB 7.-9.6.2011 Vienna

7. Kuhn,E.: Energetics of the time dependent flow behaviour of greases. Applied Rheology June 1997, p.118-122

8. Czarny,R.: Lubricating greases. *WNT Publisher,* Warsaw, 2004 (in Polish)

9. Delgado,M.A.: Manufacture and flow process of lubricating greases. PhD thesis.*University of Huelva,* Spain, 2005

10. Åström, H.: Grease in elastohydrodynamic lubrication. PhD thesis, *Lulea University,* 1993

11. Franco, J.M.;Delgado,M.A. and Valencia,C.: Combined oxidative-shear resistance of castor oil-based lubricating greases. *3rd Arnold Tross Colloquium. Hamburg 2007*. Proc. pp.18-59

12. Dierich,P.: Modelling the influence of roughnes on the wear prediction. Habil-thesis.(in German), Zittai 1986

13. Ling,F.F.; Bryant,M.D. ang Doelling,K.L.: On irreversible thermodynamics for wear prediction. Wear 253(2002)1165-1172

14. Buldum,A.; Ciraci,S.: Atomic-scale study of dry sliding friction. Phys. Rev. B55(4) (1997) 2606-2611.

15. Aghdam,A.B., Khonsari,M.M.: On the relation between wear and entropy in dry sliding contact. Wear 270(2011) 781-790

16. Doelling, K.L.; Ling, L.L.; Bryant, M.D. and Heilman, B.P.: An experimental study of the correlation between wear and entropy flow in machinery components. *Journal of Applies Physics* . 88(2000)5

17. Abdel-Aal, H.A.: On the interdependence between kinetics of friction- released thermal energy and the transition in wear mechanisms during sliding of metallic pairs. *Wear* 253, (2002), pp. 11-12

18. Abdel-Aal, H.A.: Wear and irreversible entropy generation in dry sliding.*Annals Dunarea de Jos of Galati, Fascicle*,VIII,pp.34-44

19. Basarov, I.P.: Thermodynamic. *D.V.W.*, Berlin 1964 (in German)

20. Bryant, M.D.: Entropy and dissipative processes of friction and wear. SERBIATRIB 09, *11th Intern. Conf. on Tribology* 13.5.-15.5.2009, pp.3-8.

21. Kuhn,E.: Experimental investigations of the structural degradation of lubricating greases. GfT-conference Göttingen 2012

Chapter 6

Development of Eco-Friendly Biodegradable Biolubricant Based on Jatropha Oil

M. Shahabuddin[1], M. Rahman[1], H.H. Masjuki[1], and M.A. Kalam[1]

[1]Centre for Energy Sciences, Faculty of Engineering, University of Malaya, Kuala Lumpur, Malaysia

INTRODUCTION

Various types of lubricants are available all over the world including mineral oils, synthetic oils, re-refined oils, and vegetable oils. Most of the lubricants which are available in the market are based on mineral oil derived from petroleum oil which are not adaptable with the environment because of its toxicity and non-biodegradability [1, 2]. Unknown petroleum reserve and the increasing consumption, which made concern to use petroleum based lubricant thus, to find the alternative lubricant to meet the future demand is an important issue [3]. Therefore, vegetable oil can be played a vital role to substitute the petroleum lubricant as it possesses numerous advantage over base lubricant like renewability, environmentally friendly, biodegradability, less toxicity and so on [4-8]. It has been reported that yearly 12 million tons of lubricants waste are released to the environment [9]. However, it is very difficult to dispose it safely for the mineral oil based lubricants due its toxic and non-biodegradable nature. To reduce the dependency on petroleum fuel, legislations have been passed to use certain percentage of biofuel in many countries, such initiative also required for lubricant as well [10]. Vegetable oils are mainly triglycerides which contain three hydroxyl groups and long chain unsaturated free fatty acids attached at the hydroxyl group by ester linkages acids favors triglycerides crystallization [11, 12]. The unsaturated free fatty acid which is defined as the ratio and position of carbon-carbon double bond, one two and three double bonds of carbon chain is named as an oleic, linoleic, and linoleic fatty acid components respectively [13]. The main limitations of vegetable oil are its poor low temperature behavior, oxidation and thermal stability and gumming effect [14, 15]. These stabilities and pour point behavior can be ameliorated by transesterification. Moreover the inferior flow property does not affect much in the tropical countries. Quinchia et al. [16] stated that, improving the potentiality of bio lubricants some technical properties including available range of viscosities are need to improve. To do so, environmentally friendly viscosity modifier can be used. viscosity is the most important property for the lubricants since it determines the amount of friction that will be encountered between sliding surfaces and whether a thick enough film can be build up to avoid wear from solid-to-solid contact. Since little chance of viscosity with fluctuations in temperature is desirable to keep variations in friction at a minimum, fluid often are rated in terms of viscosity index. The less the viscosity is changed by temperature, the higher the viscosity index. Ethylene–vinyl acetate (EVA) and styrene–butadiene–styrene (SBS) copolymers were used to increase the viscosity range of high-oleic sunflower oil, in order to design new environmentally friendly lubricant formulations with increased viscosities. The maximum kinematic viscosities, at 40 and 100 °C, were increased up to around 150–250 cSt and 26–36 cSt, respectively [17].

Despite of having lot of advantages of bio lubricant over petroleum based lubricant, the attempt to formulate the bio lubricant and its applications are very few. Thus, in this article we sought to extend our investigation and to test the tribological characteristics and compatibility of non-edible Jatropha oil based bio lubricant for the automotive application. The reason of selecting Jatropha oil as a base stock is it does not contend with the food and can be grown in marginal land.

EXPERIMENTAL

Lubricant Sample Preparation

There were six different types of lubricant sample were investigated in this study. The lubricant SAE 40 was used as a base lubricant and comparison purpose. Others samples were prepared by mixing of 10%, 20%, 30%, 40% and 50% Jatropha oil in SAE 40. The samples were mixed with the base lubricant by a homogeneous mixture machine.

Friction and Wear Evaluation

The apparatus used in the friction and wear testing process were Cygnus Friction and Wear Testing Machine which is connected with a personal computer (PC) with data acquisition system. It is a tri-pin-on-disc machine which is conducted by using three pins on a disc as testing specimens. Specifications of the Cygnus Test Machine are tabulated in Table 1. The block diagram of friction and wear testing are shown in Fig. 1. During the test the load of 30N and rotational speed of 2000 rpm were applied on pin.

Table 1: Specification of Cygnus wear testing machine

Parameter	Value
Test Disc Diameter	110.0 mm
Test Pin Diameter	6.0 mm
Test Disc Speed Range	25 to 3000 rpm

Motor	Tuscan; (2000 rpm, 1.5 kW)
Load Range	0 KG to 30 KG
Electrical Input	220 Volt AC 50 Hz

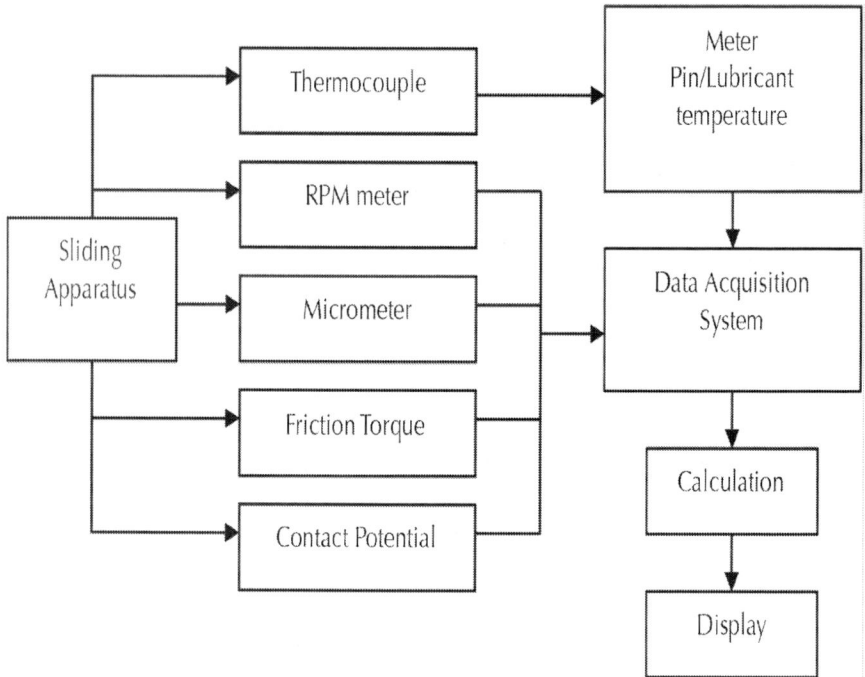

Figure 1: Block Diagrams of Friction and Wear Testing.

Preparation of the specimen

The specimens were prepared from aluminum and cast iron material. Aluminum was used to build three pin and cast iron is used for disc specimen. The construction geometry and the dimension are shown in Fig. 2. Prior to conduct the test it was ensured that the surface of the specimens are cleaned properly i. e, free from dirt and debris. Alcohol was used for cleaning purpose.

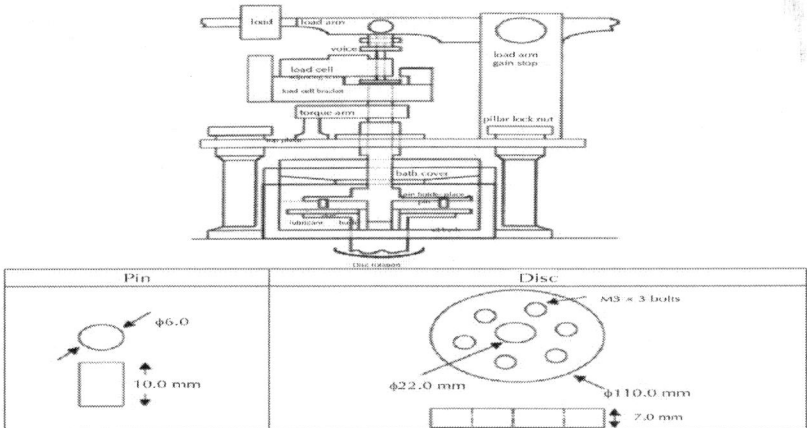

Figure 2: Schematic diagram of the experimental set up and dimensions geometry's of pins and disc specimen.

Lubricant Analyses

Multi element oil analyzer (MOA) was used to measure the wear elements in the lubricants by Atomic Emission Spectroscopy (AES). Whereas, for viscosity measurement the automatic Anton Paar viscosity meter was used with standard ASTM D 445. Viscosity was measured for both 40°C and 100°C controlled bath temperatures.

RESULTS AND DISCUSSION

Friction and Wear Characterization

Fig. 3 show the pins wear as a function of sliding time for various Jatropha oil blended bio lubricants. At the operating condition of 2000 rpm and 30 N loads, the linear pin wear varied from 0.02 to 0.05 mm. It is observed that the maximum wear occurred in the beginning of the experiment using bio lubricants. It can be seen form the Fig. 3, that the maximum wear was occurred for JBL40 while the minimum wear was observed for JBL10. The results can be attributed to the maximum ability of the JBL 10 bio lubricant film to protect metal to metal contact and keep consistency throughout the operation time while this ability is least for JBL40. It can also be seen that the rate of wear throughout the time is almost identical for the bio lubricants whereas, the reducing trend is observed for the base lubricant. At the beginning of the test, the wear rate was very fast for few minutes which are known running-in period. During this period, the asperities of the sliding surface are cut off and the contact area of the sliding surface grows to an equilibrium size. After certain period of time, equilibrium wear condition between pins and disc surface was established and thereby the wear rate became steady. It can be identified from the Fig. 3 that the bio lubricants JBL 30, JBL 40 and JBL 50 showed high wear while base lubricant, JBL 10 and JBL 20 impart low pin wear and their values are nearly same with each other.

Fig. 4 sows the loos of material from the pin for different percentage of bio lubricant samples. It seems quite clear that the loos of material from the pins are highest for 50% bio lubricant and that is least for base lubricant. It can also be interpreted that the loos of material from JBL 10 is almost similar with base lubricant and this loos of material is increasing with increasing bio lubricant percentages.

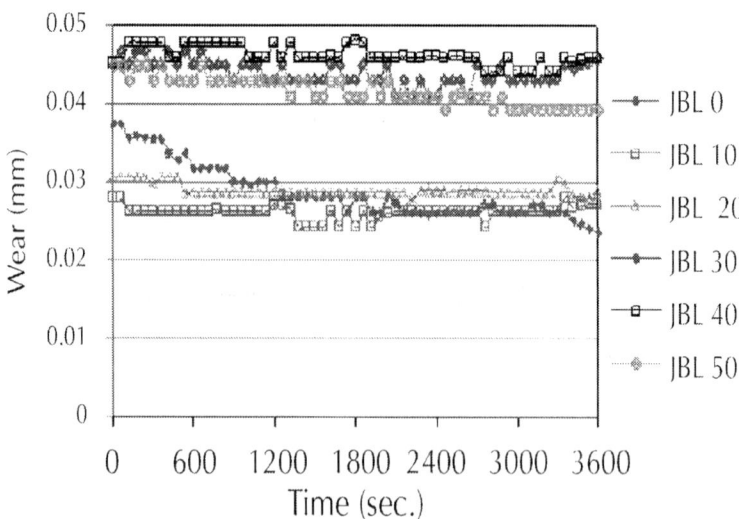

Figure 3: The linear pin wear as a function of sliding time for various Jatropha oil bio lubricants.

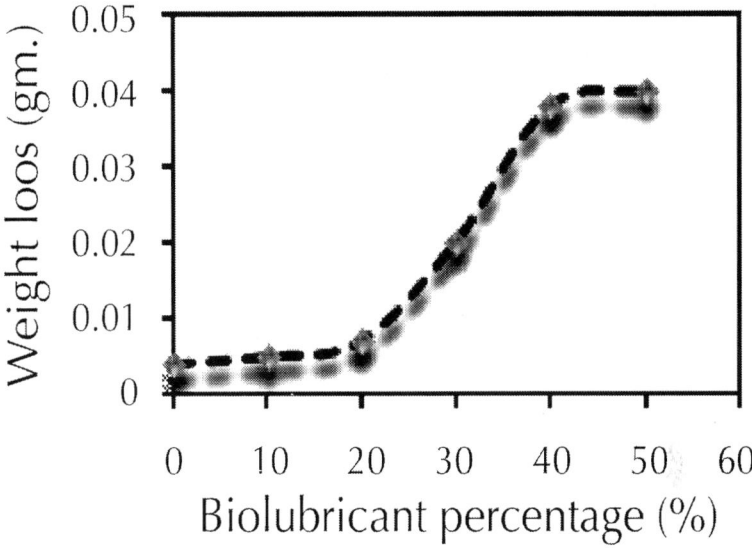

Figure 4: Loos of material form the pin for various bio lubricant percentages.

Coefficient of Friction

Fig.5 shows the friction coefficient plotted against the sliding time for various Jatropha oil bio lubricants. The results of the figure depict that the lubricant regime that occurred during the experiment were the boundary lubrication with the value of friction coefficient for boundary lubricant in the range of 0.001 to 0.2 except for 50% of Jatropha oil bio lubricant. For JBL 0, it can be seen that the coefficient of friction is highest at the beginning and then it fell down rapidly and became least with compared to all tested samples after half of the operation time. The bio lubricant percentage from 10 to 40% showed likely to be similar coefficient of friction (μ) which is almost 0.15. Whereas, the 50 % added Jatropha oil showed the coefficient of friction value of ~ 0.225 throughout the operation time. The fatty acid component of biolubricants formed multi and mono layer on the surface of the rubbing zone and make stable film to prevent the contact between the surfaces.

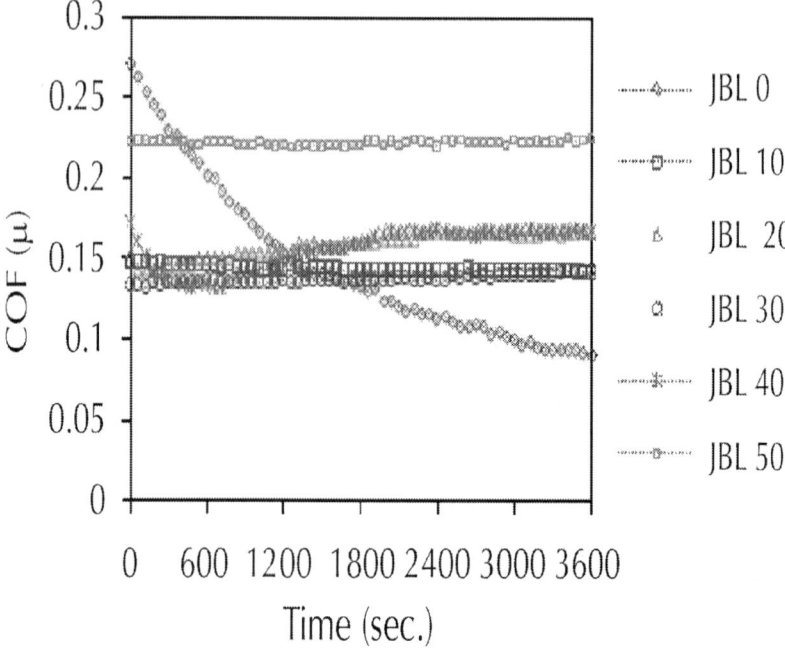

Figure 5: The Coefficient of friction as a function of sliding time for various Jatropha oil bio lubricants.

Lubricants Temperature

Fig.6 shows the relationship of the averages oil temperature of varies percentage of Jatropha oil bio lubricants with the sliding time. The rise of temperature during the running hour (1 h) for JBL 10 is least while the highest change is occurred for JBL 40 which is 11.77°c and 25.49°C respectively. The temperature rises of other samples are of 12.8°C, 18.65°C and 13. 66°C for 20% 30% and 50% Jatropha oil added bio lubricants respectively. The results of the Fig. 6 show that the JBL 10 has the highest potentiality to retain its property without much changing its temperature. From the figure it can also be interpreted that up to 30 minutes rate of change of temperature is high while the changing rate is low for second half of the operation time. It can be explained that during second half of the operation time heat produced in the lubricant due friction and the heat dissipated to the outside is nearly equilibrium.

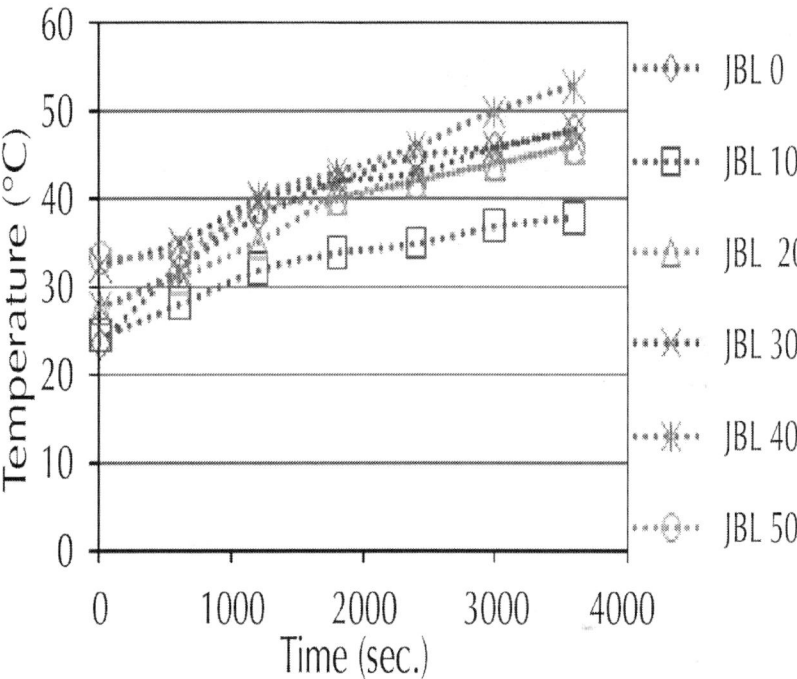

Figure 6: The Lubricant Temperature as a function of sliding time for various Jatropha oil bio lubricants.

Viscosity

Viscosity is the measure of resistance to flow [18]. Table 2 shows the viscosity grade requirement for the lubricants set by International standard organization (ISO), while Fig. 7 shows the viscosity of tested different bio lubricant samples. The comparison of the results of the Fig.7 with that of ISO grade illustrates that in case of 40°C, the bio lubricants JBL 40 and JBL 50 did not meet the ISO VG100 requirement. On the other hand all other bio lubricants meet the entire ISO grade requirement as well. It can also be noted that the viscosity of bio lubricants are much higher than standard requirements

Table 2: ISO Viscosity grade requirement [19]

Kinematic viscosity	ISO VG32	ISO VG46	ISO VG68	ISO VG100
@ 40°C	"/>28.8	"/>41.4	"/>61.4	"/>90
@ 100°C	"/>4.1	"/>4.1	"/>4.1	"/>4.1

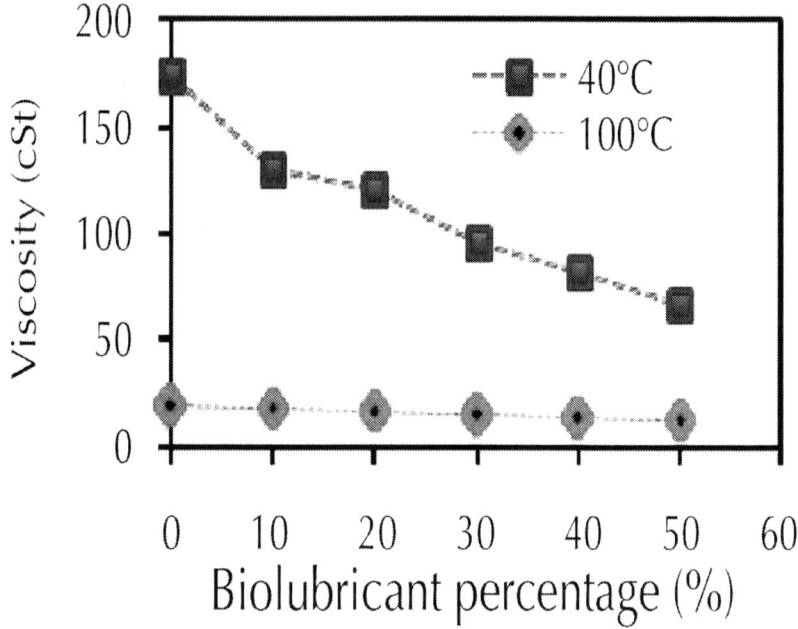

Figure 7: The viscosity of various percentages of bio lubricants at 40°C and 100°C.

Elemental Analysis

The aim of the elemental analysis by using Multi Element Oil Analyzer (MOA) is to determine the kinds and amount of metal contain in the lubricating oil. Table 3 shows the elemental analysis of tested lubricant sample by using MOA before and after the test. From the Table 3, it can be noticed that the base lubricant contains higher Silver (Ag), Zinc (Zn), Phosphorus (P), Magnesium (Mg) and Boron (B) with in high percentage compared to other element while, in pure Jatropha oil, Calcium (Ca) and Silicon (Si) are the higher element compared with other element. Some of the elements are used as additive in the lubricant to ameliorate the lubricants tribological properties. From the results, increasing number of iron (Fe) and aluminum (Al) molecules are observed with increasing percentages of Jatropha oil in the base lubricants. The source of Fe and Al are mainly cast iron plate and aluminum plate. Due to lower hardness of the aluminum pin the extraction of aluminum molecule form the pin is much higher than cast iron plate. The changes of other elements were observed before and after the test. It is clear from the elemental analysis that, most of elements were decreased after the test, by oxidizing and the chemical interaction among the elements.

Table 3: Elemental analysis of tested lubricant sample

Parameters	Types of lubricant												
Test	IBL0		IBL10		IBL20		IBL30		IBL40		IBL50		Jatropha Oil
	Before	After	Before	After	Before	After	Before	After	Before	After	Before	After	
Iron (Fe)	0.00	2.00	1.00	2.00	1.00	3.00	1.00	3.00	1.00	6.00	2.00	6.00	2
Aluminum (Al)	0.00	15.00	0.00	81.00	0.00	188.00	0.00	205	0.00	211.0	0.00	76.00	0
Copper (Cu)	0	1.00	0.00	3.00	1.00	1.00	1.00	1.00	1.00	7.00	2.00	5.00	3
Lead (Pb)	3	4.00	4.00	5.00	2.00	4.00	3.00	4.00	3.00	3.00	3.00	2.00	0
Tin (Sn)	0.00	0.00	0.00	0.00	0.00	0.00	0.00	0.00	1.00	0.00	2.00	2.00	45
Nickel (Ni)	2.00	2.00	3.00	3.00	1.00	3.00	3.00	3.00	3.00	3.00	2.00	2.00	1.5
Titanium (Ti)	0.00	0.00	1.00	1.00	0.00	1.00	0.00	1.00	0.00	0.00	1.00	1.00	1
Silver (Ag)	108	103	0.00	0.00	0.00	0.00	0.00	0.00	0.00	0.00	0.00	0.00	0
Molybdenum (Mo)	3.00	3.0	4.00	6.00	2.00	3.00	4.00	3.00	4.00	6.00	3.00	4.00	15
Zinc (Zn)	1000	771	903	716	1000	829.0	911.0	851.0	942.00	900.0	946.00	83240	1
Phosphorus (P)	500.00	428	471	441	462.00	440.0	435.00	408.0	387.00	394.0	348.00	294.00	45

Calcium (Ca)	40	30.00	37.00	33.00	35.00	27.0	28.00	21.0	23.00	29.00	21.00	17.00	18.00
Magnesium (Mg)	27	211.00	409.00	483.0	508.00	527.0	503.00	435.0	557.00	616.00	572.	637.0	748.00
Silicon (Si)	16	7.00	14.00	13.0	9.00	12.0	8.00	15.0	6.00	10.00	6.00	4.00	5.00
Sodium (Na)	4	4.00	3.00	4.00	5.00	4.00	3.00	2.0	2.00	5.00	2.00	1.00	2.00
Boron (B)	0.5	21.00	40.00	44.0	44.00	32.0	52.00	28.0	52.00	58.00	52.00	54.00	60.00
Vanadium (V)	1	1.00	1.00	0.00	1.00	0.00	1.00	0.00	0.00	1.00	0.00	1.00	0.00

Surface Texture Analysis

There are various types of wear in the mechanical system, such that abrasive wear, adhesive wear, fatigue wear and corrosive wear. Since the lubricant regime occurred in this experiment was boundary lubrication thereby, abrasive wear, adhesive wear, fatigue wear and corrosive wear were observed in to the rubbing zone. All these wears mechanisms found in this experiments but the mostly the wear phenomenon were abrasive and adhesive wear. This is because of an existence of straight grooves in the direction of the sliding direction. These grooves exist because the asperities on the hard surface (disc) touched the soft surface (pins) and hade a close relationship with the thickness of lubrication film. The optical images of the tested cast iron plate using various types of bio lubricants are shown in Fig. 8. Referring to the Fig. 8, it is found that the wear increases with increasing percentage of Jatropha oil in the bio lubricants. Reduction of lubricant film thickness leads to the surfaces to come closer to each other and cause higher wear.

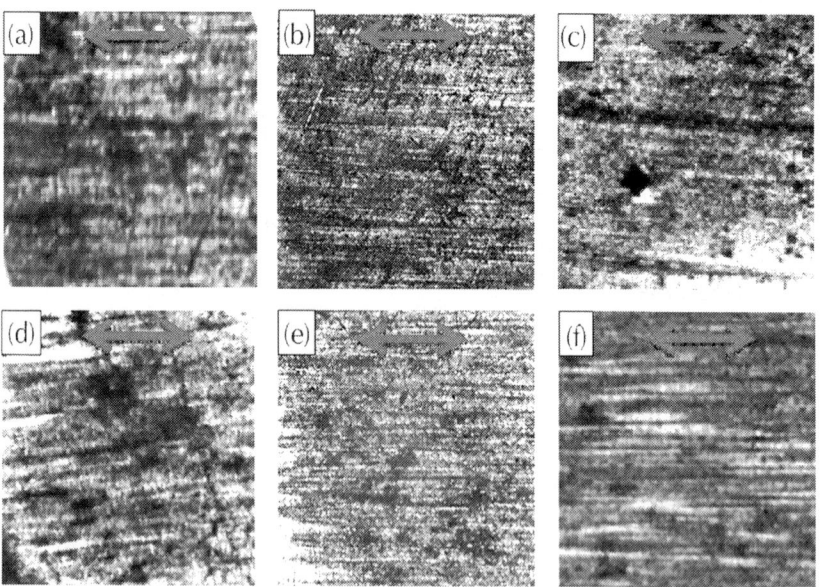

Figure 8: Optical image of the surface of the cast iron plate for different bio lubricants (magnification 30 ×): (a): JBL 0, (b): JBL10, (c): JBL 20, (d): JBL 30, (e): JBL 40, (f): JBL 50.

CONCLUSIONS

Based on the experimental study the following conclusion can be drawn:

1. The rates of wear for various percentage of bio lubricant were different. Moreover the wear rate for 10% Jatropha added bio lubricants were almost identical with base lubricant.

2. Lower the resistance to wear, higher coefficient of friction.

3. At the beginning of the test rate of wear as well as rise in temperature were high. With respect to wear rate and rise in temperature during entire operation time, the JBL 10 bio lubricant showed best performance in terms of its ability to withstand its properties.

4. From the elemental analysis of the bio lubricants, it is found, Iron and Aluminum were increased after the test due to the loos of material from the pin and the disc while, and some element like Phosphorus, Calcium and Magnesium were decreased by oxidizing and due to other chemical interaction.

5. In terms of viscosity, almost all bio lubricants met the ISO viscosity grade requirement whereas, 40% and 50% addition of Jatropha oil do not meet the ISO VG 100 requirement at 40°C.

According to the experimental result, it can be recommended that the addition of 10% Jatropha oil in the base lubricant is the optimum for the automotive application as it showed best overall performance in terms of wear, coefficient of friction, viscosity, rise in temperature etc.

ACKNOWLEDGEMENT

The authors would like to acknowledge the Department of Mechanical Engineering, University of Malaya, and Ministry of Higher Education (MOHE) of Malaysia for HIR grant (Grant No. UM.C/HIR/MOHE/ENG/07) and ERGS grant no ER022-2011A which made this study possible.

REFERENCES

1.   N. Salih, J. Salimon, E. Yousif, Synthetic biolubricant basestocks based on environmentally friendly raw materials. Journal of King Saud University-Science 2011

2.   A. Adhvaryu, Z. Liu, S. Erhan, Synthesis of novel alkoxylated triacylglycerols and their lubricant base oil properties. Industrial Crops and Products 200521113119

3.   M. Shahabuddin, H. H. Masjuki, et. Kalam, al, Effect of Additive on Performance of C.I. Engine Fuelled with Bio Diesel. Energy Procedia 20121416241629

4.   M. T. Siniawski, N. Saniei, B. Adhikari, L. A. Doezema, Influence of fatty acid composition on the tribological performance of two vegetable-based lubricants. Journal of Synthetic Lubrication 200724101110

5.   Salunkhe DK. World oilseeds: chemistry, technology, andutilization. 1992

6.   Hwang HS, Erhan SZ.Lubricant base stocks from modified soybean oil. AOCS Press: Champaign, IL; 2002

7.   T. C. Ing, A. K. M. Rafiq, S. Syahrullail, Friction Characteristic of Jatropha Oil using Fourball Tribotester. In: Regional Tribology Conference- RTC2011. Langkawi, Kedah, Malaysia: 2011

8.   M. Shahabuddin, M. A. Kalam, H. H. Masjuki, M. Mofijur, Tribological characteristics of amine phosphate and octylated/ butylated diphenylamine additives infused biolubricant. Energy Education Science and Technology Part A: Energy Science and Research 20123089102

9.   G. E. Totten, S. R. Westbrook, R. J. Shah, Fuels and Lubricants Handbook: Technology,Properties, Performance, and Testing. 2003885909p.

10.   Liaquat AM, Masjuki HH, Kalam MA et al.Application of blend fuels in a diesel engine. Energy Procedia 20121411241133

11.   N. Jayadas, K. P. Nair, Coconut oil as base oil for industrial lubricants--evaluation and modification of thermal, oxidative and low temperature properties. Tribology international 200639873878

12. N. Fox, G. Stachowiak, Vegetable oil-based lubricants-a review of oxidation. Tribology international 20074010351046

13. C. Waleska, E. W. David, C. Kraipat, M. P. Joseph, The effect of chemical structure of base fluids on antiwear effectiveness of additives. Tribol. Int. 2005383216

14. N. Ponnekanti, S. Kaul, Development of ecofriendly/biodegradable lubricants: An overview. 2012

15. M. Mofijur, H. H. Masjuki, et. Kalam, al, Palm Oil Methyl Ester and Its Emulsions Effect on Lubricant Performance and Engine Components Wear. Energy Procedia 20121417481753

16. L. Quinchia, M. Delgado, C. Valencia, et al. Viscosity modification of different vegetable oils with EVA copolymer for lubricant applications. Industrial Crops and Products 201032607612

17. L. Quinchia, M. Delgado, C. Valencia, et al. Viscosity modification of high-oleic sunflower oil with polymeric additives for the design of new biolubricant formulations. Environmental science & technology 20094320602065

18. M. Shahabuddin, Masjuki. H. H. Kalam, et al. An experimental investigation into biodiesel stability by means of oxidation and property determination. Energy 2012

19. Rudnick LR. Automotives Gear Lubricants, Synthetics, mineral oils, and bio-based lubricants: chemistry and technology. Taylor and Francis, Florida; 2006

Chapter 7

Application of Re-Refined Used Lubricating Oil As Base Oil For the Formulation of Oil Based Drilling Mud - A Comparative Study

Oghenejoboh K. M.[1*], Ohimor E. O.[2], and Olayebi O.[3]

[1]Department of Chemical Engineering, Delta State University, Oleh Campus, P. M. B. 22, Oleh, Nigeria

[2]Industries Department, Ministry of Commerce and Industry, Agbor, Delta State, Nigeria

[3]Department of Chemical Engineering, Faculty of Engineering, Federal University of Petroleum Resources, Ugbomoro, Delta State, Nigeria

ABSTRACT

The viability of using re-refined used lubricating oil (RULO) extracted with an aromatic selective normal methylpyrolidone (NMP) as base oil for the formulation of drilling mud was investigated. The rheological and other properties of this formulation were compared with formulations from three locally produced synthetic base oils. The synthetic base oils were Paradril® made from saturated linear ethylene polymer, Emcaid® manufactured from a blend of olefin isomers and Ty-Chem-Low Tox® made from catalytic dimerization of linear alpaolefins. RULO based mud, though alkaline in nature with a pH of 8.5 exhibits very poor filtration properties with the thickest filter cake when compared with the other formulations. It is also the least stable of the four formulations with an electrical stability (ES) of 480 volts. RULO formulation is very toxic as the cassava plant on which it was spilled survived for only 5 days compared to 15 days for Paradril®. It is therefore, not environmental friendly and may not also be cost effective as the cost of re-refining and extraction may far exceed the cost of producing synthetic base oil. RULO may not therefore be a viable alternative to existing base oils for the formulation of drilling mud.

INTRODUCTION

The oil industry in Nigeria had in the past relied too heavily on imported drilling chemicals (fluids) for her drilling operations which in turn have increased their operating cost and engendered capital flight. Different types of drilling fluids are used by the oil producing companies in Nigeria due to their onshore/offshore operational nature. These drilling fluids are water-based mud (WBM), oil-based mud (OBM) and synthetic-based mud (SBM). The type used for a particular drilling operation depends on the nature and location of the oil wells to be drilled. WBM which is made from bentonite clay with some chemicals such as potassium formate added to achieve various effects like viscosity control, shale stability, enhance drilling rate of penetration, cooling and lubricating of equipment are used mainly for drilling shallow onshore wells (Broni-Bediako and Amorin, 2010). However, oil wells are rarely shallow and sometimes complex evolving from vertical, inclined, horizontal, sub-sea to deep-sea drilling; as a result WBM becomes ineffective in accomplishing the required objective of an efficient drilling mud, therefore the use of OBM becomes imperative. OBM is a mud having a petroleum product such as diesel fuel as the base fluid. Oil-based muds are used for many reasons, some being ability to withstand greater heat without breaking down and cost environmental considerations. Other advantages of OBM over WBM are its excellent fluid loss control, no shale swelling, adequate lubrication of drill bits and good cutting carrying ability. Synthetic-based fluid is a mud with synthetic oil as the base fluid. This is most often used on offshore rigs because it has the properties of an oil-based mud, but the toxicity of the fluid fumes are much less than an oil-based fluid. This is important when men work with the fluid in an enclosed space such as an offshore drilling rig. Due to the environmental advantages of synthetic-based mud, it is more preferable by drilling companies despite its exorbitant cost. Synthetic oil based mud (SOBM) is basically water-in-oil or 'invert', emulsion. The water-in-oil emulsion itself is usually stabilized with a "primary emulsifier" (often a fatty acid salt), while the weighting material, along with drill solids which the mud acquires in use, is made oil-wet and dispersed in the mud with a "secondary emulsifier" (typically a strong wetting agent, such as a polyamide) (Broni-Bediako and Amorin, 2010; Growcock et al., 1994). For SOBM to be effective the emulsion formed must be stable, such stability is derived from the strong visco-elastic characteristics caused by the presence of asphaltenes and resins in the mixture (Akpabio and Ekott, 2013; Langevin et al., 2004). The SOBM premixes or invert emulsions are formulated to contain some amount of water (up to 30%). The amount of synthetic oil and water in the SOBM premix is referred to as the Oil-Water-Ratio (OWR). Emulsifiers are then added to emulsify the water as the internal phase and prevent the water from breaking out and coalescing into larger droplets (Huda and Nour, 2011). These water droplets, if not tightly emulsified, can water-wet the already oil-wet solids and dramatically affect the emulsion stability (Abdel-Raouf, 2011). To achieve this therefore, compounds with higher solubility in the oil phase rather than in the aqueous phase are used as emulsifiers (Dimitrov et al., 201).

To minimize the drilling industries' operating cost index, concerted efforts are ongoing to find an effective, inexpensive and ecologically safe drilling fluids that can be sourced locally in line with the current Nigerian Oil and Gas Industry Content Development Policy. To this end, the company, Skyward Resources Ltd based in PortHarcourt, Nigeria has developed some drilling fluids chemicals such as oil mud thinner (OMT 5), oil mud wetter (OMW 5), drilling detergent (DD 3100), primary emulsifier (PEM 5) and secondary emulsifier (SEM 5) from vegetable extracts. These chemicals have been used with biodegradable plant based oil such as jatropha oil, rapeseed oil, soyabeans oil and cottonseed oil (Fadairo et al., 2012) as well as other low aromatic synthetic mineral base oil for the formulation of SOBM that are environmental friendly and non-toxic. However, it is the belief of the authors that the production cost of drilling fluids can further be reduced by using discarded used lubricating oil as base oil for the formulation of SOBM, since plant oil is not usually available in commercial quantity. This is however, based on the fact that the used lubricating oil must meet the basic environmental requirements for such use (Nweke and Okpokwasili, 2003). Used lubricating oil is currently a source of environmental nuisance in Nigeria since it is indiscriminately dumped into rivers, soil and the environment as a result of lack of stringent enforcement of environmental laws (Oghenejoboh and Ohimor, 2012; Ogbo et al., 2009). Used lubricating oils can therefore be collected at no cost from mechanic workshops and other outlets involved in rotating machine repairs and maintenance. Used lubricating oil contains lot of impurities such as mixture of high molecular weight aliphatic and aromatic hydrocarbons as well as heavy metals acquired from engine wear and tear (Wang et al., 2000). Used lubricating oil also contains combustion products (water, un-burnt fuel, soot and carbon) as well as abrasive materials such as road dust. All these contaminants must be removed through re-refining before it can be used as base oil for the formulation of drilling mud. Re-refining of used lubricating oil involve three steps - dehydration, stripping and distillation. The dehydration step entails physical treatment in which the used oil is stored in a container for a period of time to allow water and solids to separate out of the oil followed by boiling to break water emulsion and to allow fuel diluents to evaporate from the oil. The stripping step involves normal fractionation where the bulk of the feedstock is distilled off as lubricating oil fractions. The final step in the re-refining process is the extraction process whereby a suitable solvent is used to remove all carcinogenic compounds such as poly aromatic compounds contained in the oil. This step also remove odour and colour from the oil. In the present study, the stability and toxicity of SOBM formulated from re-refined used lubricating oil is compared with those from three commercial base synthetic base oil - Paradril® (made from saturated linear ethylene polymer), Emcaid® (made from a blend of olefin isomers) and Ty-Chem-Low Tox® (made from catalytic dimerization of linear alphaolefins) as well as results obtained with plant base oil by other workers.

MATERIALS AND METHODS

The materials used for the experiments were spent lubricating oil, soxhlet extractor fixed with 500 ml flask, distillation column, digital weighing balance, Hamilton beach mixer, mud balance, hot plate, digital thermometer, 1000 ml measuring cylinders, 500 ml measuring cylinder, 100 ml beaker, 5 ml syringes and ES-meter. Other materials used were, synthetic base oils, primary emulsifier (PEM 5) and secondary emulsifier (SEM 5) both obtained from Skyward Resources Ltd based in Port-Harcourt, organoclay, soltex, lime, calcium chloride, distilled water and barite.

Table 1: Viscometer reading for mud formulated from the base oils used in this work

Dial reading (D) (RPM)	Base oil samples (lb/100 ft^2)			
	A	B	C	D
600	186	122	144	130
300	168	111	129	109
100	158	93	124	101
100	151	86	114	92
60	147	71	105	85
30	136	68	93	70
3	72	51	68	55

Experimental Procedure

Treatment and Re-Refining of Used Lubricating Oil

Ten litres of used lubricating oil obtained from a motor mechanical workshop in Warri, Delta State of Nigeria was left in a 20 L plastic paint bucket for 5 days to allow water and solids to separate out of the oil after which the oil was decanted. Some of the decanted oil was then heated in a closed vessel immersed in a water bath maintained at 120°C for 60 min to boil off some of the emulsified water and fuel diluents. The dehydrated oil was then fractionated using a laboratory scale distillation column following the normal crude oil distillation process. The refined lubricating oil obtained as intermediate from the fractionation process is then extracted with Nmethylpyrolidone (NMP) using a soxhlet extractor. The extraction step is aimed at removing unwanted aromatic contaminants present in the paraffinic lubricating oil fraction since NMP is an aromatic selective solvent. The solvent also removes color and odor from the oil. The re-refined lubricating oil was then used as base oil for the formulation of drilling mud.

Formulation of Drilling Mud

175 ml of the re-refined lubricating oil and 75 ml of de-ionized water were measured into a mixing vessel using the measuring beakers. 4, 6, 6 and 2 g of organophilic clay, lime, PEM 5 and SEM 5 were then added to the mixture. 0.5 ml of brine solution prepared from 25 g of $CaCl_2$ in 100 ml of de-ionized water was added before subjecting the mixture to thorough mixing using Hamilton Beach mechanical mixer, model 936 to attain a homogeneous mixture. The formulated drilling mud was allowed to age for 24 h. The same procedure was repeated for the three synthetic base oils (Paradril®, Emcaid® and Ty-Chem-Low Tox®). For ease of identification, the base oil samples used for the drilling fluid formulation were labeled:

Sample A: re-refined used lubricating oil

Sample B: Paradril® synthetic base oil

Sample C: Emcaid® synthetic base oil

Sample D: Ty-Chem-Low Tox® synthetic base oil

Measurement of Formulated Fluid Properties

The density, viscosity, gel strength, pH, filtered volume, filter cake thickness, electrical stability as well as the toxicity of the formulated drilling fluids were determined and compared. The density and viscosity of the fluids were measured using the method outlined by Fadairo et al. (2012) with the values of apparent viscosity (μ_A), and plastic viscosity (μ_P) obtained from the equations developed by Amorin et al. (2011) as reproduced below.

$$\mu_P = D_{600} - D_{300} \tag{1}$$

$$\mu_A = \frac{D_{600}}{2} \tag{2}$$

Where D_{600} and D_{300} is the viscometer dial reading at 600 and 300 rpm in centipoises (cP) respectively.

The electrical stability of the tested drilling fluids was determined using an ES meter according to API 13B-2 procedure. Gel strength was determined using the rotational viscometer at 10 s and 10 min respectively, while the pH of the fluids was estimated by means of the pH colorimeter paper method of Fadairo et al. (2012). The API filter press was used to determine the filtered volume of the drilling fluid following the procedure of Amorin et al. (2011). The filter cake thickness of the fluids was determined using the filter paper and cake formed during the filtered volume experiment. The filter paper was thoroughly washed and placed in between two glass slides of equal diameter as the filtered paper before subjecting it to a pressure of 300 N/m2 for 3 min. Then the slides, the filter paper and cake formed were put in an extensometer to determine the thickness of the cake formed.

To test the environmental friendliness of all the formulated fluids, 100 ml of each were spilled on 4 weeks old cassava plants and the number of days of the plants' survival was noted.

RESULTS

The result of the viscosity test is presented in Table 1, while Table 2 shows the pH, density, plastic viscosity, apparent viscosity, gel strength (10 s/10 mins) as well as the electrical conductivity of the formulated fluids.

DISCUSSION

From the results presented in Table 2 we can see that sample A (re-refined used lubricating base oil) has the highest apparent viscosity followed by sample C (Emcaid® synthetic base oil) while sample B (Paradril® synthetic base oil) exhibited the least viscosity. This result infer that rerefined lubricating base oil offers the greatest resistance to fluid flow with the least resistance offered by Paradril® synthetic base oil. Re-refined used lubricating base oil therefore posed the least prospect for the formulation of a good drilling fluid when compared with the three synthetic base oils used in this work since low viscosity drilling fluid lead to reduced wear in the drill string (Mitchell, 1995). However, the formulated muds from the four base oils have similar rheological behavior as they all approximately exhibit the Bingham plastic model from the plots of the rotary viscometer dial reading against speed generated as shown in Figure 1. This is an indication that re-refined used lubricating oil has the potential to be used as base oil for formulating drilling mud if the viscosity is reduced by adding appropriate polymers. The formulated drilling fluid from the four base oil samples show the same range of densities, with Ty-Chem-Low Tox® synthetic base oil having the highest density of 8.47 ppg followed by re-refined used lubricating oil (8.32 ppg) and Paradril® with 8.30 ppg while Emcaid® had the least density of 8.13 ppg. According to Fadairo et al. (2012) the denser the base oil, the higher the amount of barite needed to build. From the results it is evident that Ty-Chem-Low Tox® and re-refined used lubricating oil that have slightly higher densities will require the highest amount of barite to build.

Table 2: Rheological and other properties of the formulated drilling muds

Base oil samples	pH	Density	Plastic viscosity (μ_p)	Apparent viscosity (μ_A)	Gel strength	ES
	(-)	(ppg)	(cp)	(cP)	lb/100 ft²	(volts)
A	8.5	8.32	18	93	53/54	480
B	8.8	8.30	11	61	55/55	697
C	9.8	8.13	15	72	60/72	550
D	7.7	8.47	21	65	48/42	596

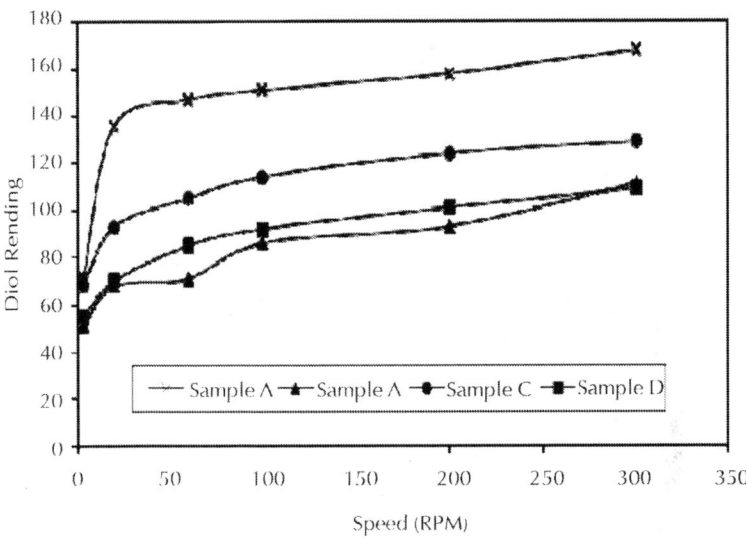

Figure 1: Viscometer plot for the formulated drilling muds.

Hydrogen ion potential (pH) is a very important parameter to consider when formulating drilling mud. Effective drilling muds are expected to be highly alkaline (that is, pH >7). This is because acidic (low pH) mud increases the corrosion of metals (pipes and casing) when it comes in contact with it. A drilling mud having a pH of between 7 and 9.5 had been reported to have the least effect on bentonite since the viscosity of such fluid remains relatively constant over a wide range of temperatures (Fadairo et al., 2012). However, a pH above 9.5 increases the mud viscosity thereby affecting the effectiveness of the drilling mud leading to complicated shale problems. As we can see from Table 2 the pH of the four formulated drilling fluids fall within the desired value, however, fluids formulated from re-refined used lubricating oil and Paradril® appear to give best hole stability and control over mud properties, since these requirements are met by fluid having a pH of 8.5 to 9.5 (Fadairo et al., 2012).

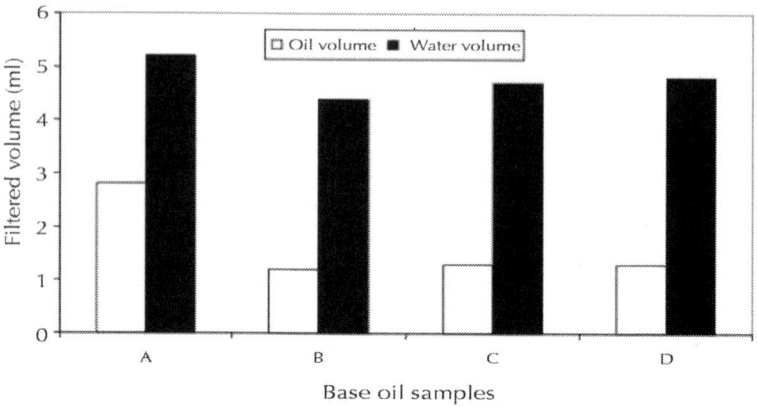

Figure 2: Filtration property of formulated drilling muds.

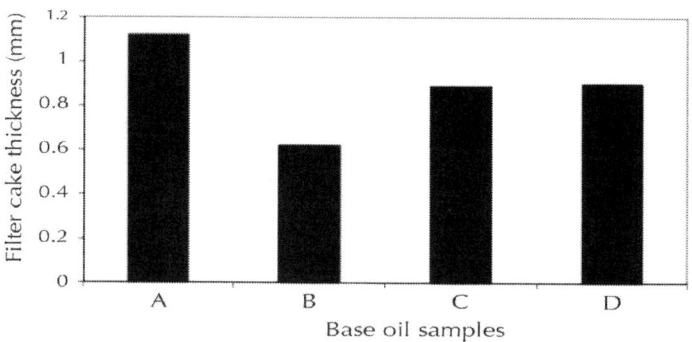

Figure 3: Filter Cake Thickness of formulated drilling muds.

The gel strength of the drilling fluids formulated from rerefined used lubricating oil was compared with those formulated from the three synthetic base oil used in this work. Gel strength is the ability of a drilling mud to suspend cuttings and other solid additives. From Table 2, the gel strength of mud produced from re-refined used lubricating oil and Paradril® synthetic base oil promised to be more effective than mud formulated from the other two synthetic base oils, since the shear rate of the mud remained consistent and high. High gel strength mud has the ability to suspend drill cuttings along the length of the drillpipe or bore annulus when the drilling mud circulation is stopped during pump tripping or any other secondary operations (Shah et al., 2010). A low gel strength mud on the other hand do not efficiently suspend cuttings thereby allowing cuttings to quickly drop leading to pump shutdown, stuck pipe, hole pack-off, barite sag as well as accumulation of cutting beds. From the results, it is clear that re-refined used lubricating oil mud has excellent cutting transport capabilities even at low values of viscosity. This result is similar to that obtained by Fadairo et al. (2012) for diesel based mud.

Another factor determining the successful performance of a drilling fluid tested for, in the formulated muds was the mud filtration capacity. From Figures 2 and 3 we can see that re-refined OBM has the highest filtration rate and as a result a thicker filter cake due to its high porosity while Paradril® SOBM exhibited the lowest rate of filtration with thinner filter cake. High filtered volume is associated with thick filter cake because the cake is formed by deposition of clay particles on the walls of the hole during loss of water to the formation. So the higher the filtered volume, the thicker the filter cake and the less efficient the drilling mud. A thick cake reduces the effective diameter of the hole and increases the contact area between the tube and the cake leading to increased risk of stuck tubes (Amorin et al., 2011). Based on this result, drilling mud formulated from re-refined used lubricating oil will not be an effective fluid for drilling purposes.

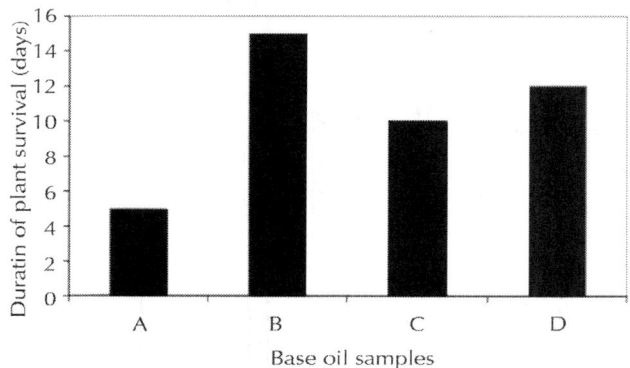

Figure 4: Survival days of cassava plants spilled with equal volume of the formulated muds.

Electrical stability (ES) is a vital property of oil based mud (OBM) and synthetic oil based mud (SOBM). The ES represents the stability of emulsions formed by oil and water during the formulation. A low ES mud is not conductive and therefore cannot transfer power. A good drilling mud should have an ES of between 700 and 900 V under circulation. However, an ES range of 300 to 400V is considered ideal for newly formulated mud as well as mud in storage. From the results of the ES test presented in Table 2, drilling mud formulated from the four base oils used in this study meet the specification for stable fluid, however, mud formulated from re-refined used lubricating oil exhibited the least ES value of 480 V and as such is the least stable of the formulations. The low ES value of the re-refined used lubricating oil based mud may be as a result of the low resistivity of the rerefined used oil. The resistivity of this base oil may have been reduced due to the rigorous re-processing steps it was subjected to prior to its use for the formulation. For re-refined used lubricating oil based mud emulsion to be stable there will be need to add water and salt to the formulation, which will invariably affect the effectiveness of the mud.

The toxicity test conducted by spilling equal volume of the four formulated muds on young cassava plants show that re-refined used lubricating oil mud is the most toxic of the formulations. Cassava plant spilled with re-refined used lubricating oil based mud first showed evidence of withering after 3 days and finally died after 5 days. Cassava plants spilled with mud formulated from Paradril® , Emcaid® and Ty-Chem-Low Tox® synthetic base oils survived for 15, 10 and 12 days respectively (Figure 4). From this result it is clear that SOBM is more environmental friendly than re-refined used lubricating oil based mud. In a similar study, Fadairo et al. (2012) observed that jatropha oil based mud spilled on growing bean seedling was able to survive for 16 days before it eventually died while the same quantity of diesel oil based mud spilled on the same bean seedling survived for only 7 days before dying. Re-refined used lubricating oil based mud is therefore more toxic than all other types of base oil used for formulating drilling mud – even diesel. Toxicity of drilling mud is a function of the aromatic content of the base oil. An environmental friendly drilling mud is one with negligible carcinogenic poly-aromatic compounds. This explains why vegetable base oil mud are ecologically and environmentally friendly as seen from the drilling mud produced from the other three synthetic base oils. Though extraction of the re-refined used lubricating oil with an aromatic selective solvent (Nmethylpyrolidone (NMP)) is aimed at reducing the aromatic content of the re-refined oil to non-detectable level, the toxicity result shows that the re-refined used lubricating oil based mud may still contain high concentration of aromatic compounds and this may have been responsible for its high toxic nature. Based on this result, re-refined used lubricating oil does not meet the environmental conditions for the formulation of an efficient and ecologically safe oil based drilling mud.

CONCLUSIONS

The possibility of using re-refined used lubricating oil as base oil for the formulation of drilling mud had been investigated. From the results, it is clear that re-refined used lubricating oil is not a viable option neither for diesel oil based mud nor for synthetic oil based mud. Re-refined used lubricating oil based mud is very toxic and therefore fails the environmental requirement as outlined for efficient drilling mud by the Nigerian Government. The cost index for re-refined used lubricating oil based mud may also be higher than those of the synthetic oil based mud due to the combined cost of refining and extraction. As a result, re-refined used lubricating oil may not be a viable alternative to vegetable oil and other synthetic oils for the formulation of drilling mud.

REFERENCES

1.   Abdel-Raouf ME (2011). Factors Affecting the Stability of Crude Oil Emulsions, www.intechopen.com. Assessed 3rd November 2-12. pp. 188.

2.   Akpabio EJ, Ekott EJ (2013). Application of Physico-Technological Principles in Demulsification of Water-In-Crude Oil System. India J. Sci. Technol. 6(1):1-3.

3.   Amorin LV, Nascimento RCA, Lira DS, Magalhães J (2011). Evaluation of the Behavior of Biodegradable Lubricants in the Differential Sticking Coefficient of Water Based Drilling Fluids. Braz. J. Pet. Gas. 5(4):197-203.

4.   Broni-Bediako E, Amorin A (2010). Effects of Drilling Fluid Exposure to Oil and Gas Workers Presented with Major Areas of

Exposure and Exposure Indicators. Res. J. Appl. Sci. Eng. Tech. 2(8):770-772.

5.   Dimitrov AN, Yordanov DI, Petkov PS (2011). Study on the Effects of Demulsifiers on Crude Oil and Petroleum Products. Int. J. Environ. Res. 6(2):435-436.

6.   Fadairo A, Falode O, Ako C, Adeyemi A, Ameloko A (2012). Novel Formulation of Environmentally Friendly Oil Based Drilling Mud, In New Technologies in the Oil and Gas Industry, Chapter 3, INTECH Open Science, http://dx.doi.org/105772/52136 Assessed 4th January 2013.

7.   Growcock FB, Ellis CF, Schmidt DD (1994). Electrical Stability, Emulsion Stability, and Wettability of Invert Oil-Based Muds. SPE Drilling Completion 9(1):39-46.

8.   Huda SN, Nour AH (2011). Microwave Separation of Water-in-Crude Oil Emulsions. Int. J. Chem. Environ. Eng. 2(1):70-71.

9.   Langevin D, Pateau S, Hénaut I, Argillier JF (2004). Crude Oil Emulsion Properties and their Application to Heavy Oil Transportation. Oil Gas Sci. Technol.= Rev. IFP. 59(5):513.

10.   Mitchell B (1995). Advanced Oil Well Drilling Engineering Handbook, Mitchell Engineering, and 10th Edition. pp. 248-251.

11.   Nweke CO, Okpokwasili GC (2003). Drilling Fluid Base Oil Biodegradation Potential of a Soil Staphylococcus Species, Afr. J. Biotechnol. 2(9):293.

12.   Ogbo EM, Avwerovwe U, Odogu G (2009). Screening of four common weeds for use in phytoremediation of soil contaminated with spent lubricating oil. Afr. J. Plant Sci. 3(5):102.

13.   Oghenejoboh KM, Ohimor OE (2012). Contamination of Soil and Rivers from Used Engine Oil: A Case Study of Choba Community in PortHarcourt, Nigeria. Pol. Res. 32(2):131.

14.   Shah SN, Narayan PE, Shanker H, Ogugbue CC (2010). Future Challenges of Drilling Fluids and Their Rheological Measurements, American Association of Drilling Engineers (AADE) Conference and Exhibition, Houston Texas, USA, 6 - 7 April.

15.   Wang QR, Cui YS, Liu XM, Dong YT, Christie P (2000). Soil Contamination and Plant Uptake of Heavy Metals at Polluted Sites in China. J. Environ. Sc. Health Part A - Toxic/Hazardous Substances and Environmental Engineering 8:823-825.

Composition-Explicit Distillation Curves of Waste Lubricant Oils and Resourced Crude Oil: A Diagnostic for Re-Refining and Evaluation

[2]Lisa Starkey Ott, [1]Beverly L. Smith, and [2]Thomas J Bruno

[1]Physical and Chemical Properties Division, National Institute of Standards and Technology, Boulder, CO

[2]Department of Chemistry and Biochemistry, California State University, Chico, Chico, CA 95929-0210

ABSTRACT

Problem Statement

We have recently introduced several important improvements in the measurement of distillation curves for complex fluids. The modifications include a composition explicit data channel for each distillate fraction and temperature measurements that are true thermodynamic state points that can be modeled with an equation of state. The composition-explicit information is achieved with a sampling approach that allows precise qualitative as well as quantitative analyses of each fraction, on the fly. We have applied the method (called the advanced distillation curve technique) to a variety of fluids, including simple n-alkanes, rocket propellants, gasoline, jet fuels, diesel and biodiesel fuels and crude oils (both petroleum-and bio-derived crude oils).

Approach

In this study, we present the application of the method to new, recycled and resourced heavy oils. The ultimate purpose of this work is waste reduction and energy utilization, by placing the reprocessing steps on a more fundamental footing. First, we present measurements on four unused automotive crankcase oils and then four samples of used oils: automotive oil, cutting oil, transformer oil and a commingled lubricant waste stream. Using the advanced distillation curve metrology, we can distinguish between the different weights (viscosity ranges) of crankcase oils and compare them to the sample of used crankcase oil. The distillation curves also provide valuable information regarding the presence or absence of low-boiling contaminants in the recycled automotive oil, such as water and gasoline.

Results

Additionally, we demonstrate the evaluation of all four used lubricant oils. Then, we apply the advanced distillation curve method to a sample of crude oil prepared using a plastic waste stream from an automotive plant.

Conclusion

Overall, we conclude that the composition-explicit advanced distillation curve metrology is important for understanding the boiling behavior of the various oils streams and is a valuable diagnostic for future re-refining of the used lubricant oils. This information will be essential in the development of models for the thermophysical properties of such fluids.

INTRODUCTION

Lubricant Oils

In the United States alone, over two hundred million gallons (approximately 7.6×108 L) of used lubrication oil are disposed of improperly each year (American Petroleum Institute, 2009). The oil from one improperly disposed automotive oil change (4-5 L) can contaminate one million gallons of fresh water. It is quite clear, then, that proper collection of used lubricant oils is of great environmental importance. The next question becomes what to do with these used oils once they have been collected.

The majority of used lubricant oil that is generated is automotive crankcase oil (sometimes called drain oil). This oil can be recycled and treated for subsequent use as automotive oil, or used for other applications. The process is called re-refining and is a multi-step, complex and expensive process (OilRe-refining.aspx, 2009). First, the lightweight impurities are removed, then middle-weight impurities (such as components of gasoline) are removed via fuel stripping (Bruno, 1994). Next, vacuum distillation is employed to separate the heaviest impurities. The cleaned oil is then hydrotreated, followed by the addition of an appropriate additive package to complete the re-refining process. The totality of these individual steps results in a very energy intensive set of tasks and any improvements in either the process or product are critical to ensuring the viability and economic acceptability of re-refining.

The re-refined oil must meet the same specifications as unused crankcase oil in order for the re-refined product to be of value. There are a number of standards and specifications that are used to ensure the quality of lubricating oils; among them is ASTM Method D-4485 (Orbital Engine Company, 2003). In this standard, several other ASTM methods are used and incorporated by reference; these include, for example, flash and fire point (ASTM D-92), corrosivity (ASTM D-130) and foaming tendency (ASTM D-892). The boiling behavior (i.e., the distillation curve) of the oil is typically determined by ASTM D-2887, a gas chromatographic method that calculates the distillation curve boiling temperatures based on standardized retention times (ASTM Standard D2887-02, 2004). The distillation curve is one of the most important properties that can be measured for any complex fluid, since it is the only practical avenue to assess the volatility or the Vapor Liquid Equilibrium (VLE) (Kister, 1988; 1991; Manovyan et al., 1983) Moreover, it can be directly correlated to many engine operating parameters (Cherepitsa et al., 2003; Emel'yanov et al., 1981; 1982; Shin, 1997; Sjogren et al., 1996; Hallett and Ricard, 1992; Karonis et al., 1998; 2003; ASTM Standard D 86-04b, 2004). The D-2887 method has a stated uncertainty of up to 11.8°C in the determination of the boiling points near the end of the distillation curve. This large uncertainty in temperature and the ambiguity in the pressure at which the measurement is performed, makes the boiling behavior for crankcase oil calculated with D-2887 unsuitable for fundamental applications such as equation of state development. This disconnect with theory is a major disadvantage, because all process design simulations require an equation of state for the fluid being treated, even if that fluid is a highly complex mixture.

Consequently, a method for measuring the distillation curve of automotive oils, lubricant oils and other heavy oils with a much lower uncertainty is critical. Reduced pressure distillations, also known as vacuum distillations, are one possibility; however, the considerable uncertainty in measuring and maintaining the reduced pressures make vacuum distillations unsuitable for measurements required in equation of state development. This makes an atmospheric distillation method desirable for providing temperature and pressure measurements with low uncertainty.

Advanced Distillation Curve Metrology

The most common presentation of an atmospheric pressure distillation curve is a plot of the boiling temperature against volume fraction. The standard test method, ASTM D-86, provides the usual approach to measurement (ASTM Standard D 86-04b, 2004). The data obtained with ASTM D-86 are the initial boiling point, the temperature at predetermined distillate volume fractions and the final boiling point. The ASTM D-86 test suffers from many drawbacks, including large uncertainties in temperature measurements and little theoretical significance. Indeed, the major utility of the method stems from its standardization; everyone does it the same way.

In an effort to remedy the shortcomings of the standard distillation method described above, we have recently reported in detail an improved distillation method and apparatus. Improvements to the traditional distillation apparatus include (1) a composition-explicit data channel for each distillate fraction (for both qualitative and quantitative analysis), (2) temperature measurements that are true thermodynamic state points that can be modeled with an equation of state, (3) temperature, volume and pressure measurements of low uncertainty suitable for equation of state development, (4) consistency with a century of historical data, (5) an assessment of the energy content of each distillate fraction, (6) qualitative, quantitative and trace chemical analysis of each distillate fraction, (7) assessment of the corrosivity and stability of each distillate fraction. The new method is called the composition-explicit or advanced distillation curve technique (ADC) (Bruno, 2006a; 2006b; Bruno and Smith, 2006a; 2006b; Ott and Bruno, 2007a; 2007b; Smith and Bruno, 2006; 2007a; 2007b). A schematic diagram of the instrument is presented in Fig. 1. We have applied this method to a wide variety of gasoline, diesel fuel, aviation fuel and rocket propellant mixtures (Bruno and Smith, 2006b; Bruno et al., 2009a; 2009b; 2009c; 2009d; Hadler et al., 2009; Ott et al., 2008a; 2008b; 2008c; 2008d; Ott and Bruno, 2008; Smith and Bruno, 2007c; 2008; Smith et al., 2008a; 2008b; Lovestead and Bruno, 2009a). Moreover, we have used the measurements derived from this method as the basis for the development of thermophysical property models for many of these fluids (Huber et al., 2008a; 2008b; 2009a; 2009b).

Referring to Fig. 1, the stirred distillation flask is placed in an aluminum heating jacket contoured to fit the flask. The jacket is resistively heated, controlled by a model predictive PID controller that applies a precise thermal profile to the fluid. Three observation ports are provided in the insulation to allow penetration with a flexible, illuminated bore scope.

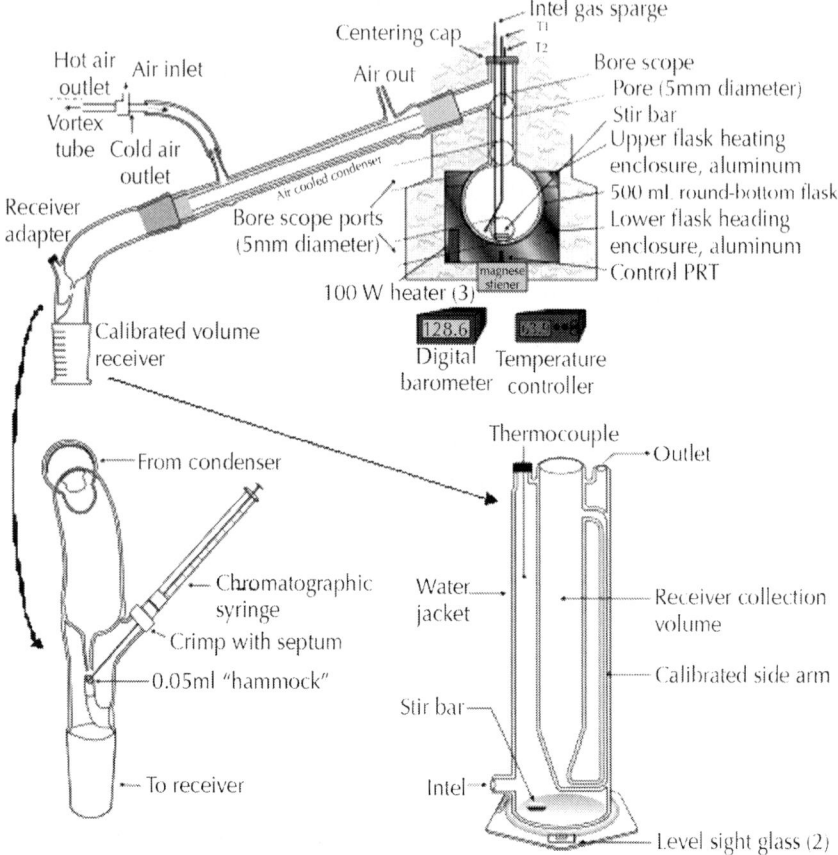

Figure 1: Schematic diagram of the overall apparatus used for the measurement of distillation curves. Expanded views of the sampling adapter and the stabilized receiver are shown in the lower half of the figure.

The ports are placed to observe the fluid in the boiling flask, the top of the boiling flask and the distillation head (at the bottom of the take-off). Above the distillation flask, a centering adapter provides access for two thermally tempered, calibrated thermocouples that enter the distillation head. One thermocouple (T1) is submerged in the fluid and the other (T2) is centered at the low point of distillate take-off. Also in the head is an inert gas blanket for use with thermally-unstable fluids. Distillate is taken off the flask with a distillation head, into a forced-air condenser chilled with a vortex tube. Following the condenser, the distillate enters a new transfer adapter that allows instantaneous sampling (on a "hammock" of known volume) of distillate for analysis. When the sample leaves the adapter, it flows into the calibrated, level-stabilized receiver for a precise volume measurement.

For each ADC measurement, we can record a data grid consisting of: T_k, the temperature measured in the fluid (with T1), T_h, the temperature in the head (measured with T2), the corresponding fluid volume, the elapsed time and the external (atmospheric) pressure. Along with these data, one withdraws a sample for detailed analysis, which is the most important advantage presented by the ADC. Sampling very small volumes of the distillate yields a composition-explicit data channel with nearly instantaneous composition measurements. Chemical analysis of the distillate fractions allows us to understand how the composition of the fluid varies with volume fraction and distillation temperature, even for complex fluids. The fraction-by-fraction chemical analysis coupled with the distillation curve (which can be used to approximate vapor-liquid equilibrium of complex mixtures) provides a link between chemical analysis and thermodynamics.

The application of an atmospheric distillation curve measurement to heavy fluids such as lubricant oils requires some discussion. Although atmospheric distillation methods are routinely applied to heavy fluids for many diagnostic or Quality Assurance/Quality Control (QA/QC) applications, a reduced pressure method is sometimes preferred because of the lower temperatures that result. Clearly, at the higher temperatures encountered in atmospheric distillations, it is often not possible to obtain a meaningful final boiling temperature, due to sample decomposition. When approached with the ADC and its composition-explicit data channel, compositions measured from those high temperatures may indeed reveal such decomposition. The ultimate goal of our work is not QA/QC, but rather the development of reliable fluid models that depend on correlations of experimental measurements. As such, we require pressure control and measurement of very low uncertainty, not generally obtainable with reduced pressure approaches (reduced pressure measurements have temperature uncertainties up to 8°C and pressure uncertainties up to 18%) (Manovyan et al., 1983). For this reason, our measurements of heavy fluids do not generally extend to the final boiling point. Additionally, thermal cracking during the distillation can be probed with any applicable analytical technique (primarily gas chromatography-mass spectrometry, GC-MS) on the distillate fractions. Indeed, GC-MS evaluation of the distillate fractions of petroleum crude oils indicates that these heavy oils do not undergo significant thermal cracking until well above 400°C. Moreover, kinetic determinations of the thermal decomposition of complex fluids lead to a similar conclusion. For example, in studies of the thermal decomposition or cracking of the hydrocarbon fuels Rocket Propellant 1 (RP-1), Rocket propellant 2 (RP-2) and aviation fuel Jet-A, it was found that significant decomposition/cracking of the fluid did not occur until temperatures significantly above 400°C were reached (Lovestead and Bruno, 2009a; 2009b; Huber *et al.*, 2008a; 2008b). Moreover, these measurements were done at a pressure of 35 MPa (approximately 5000 psi).

While we recognize the difference between a heterogeneous mixture such as a multi-grade oil with detergents and viscosity improving additives and a finished fuel such as RP-1, there is nonetheless a great deal of similarity in the chemical behavior especially among the nonpolar moieties. Given appropriate constraints and cautions, we find that the ADC is a useful tool for measuring the atmospheric distillation curve (and composition by distillate fraction) of heavier complex fluids such as oils.

Crude Oil Prepared from Waste

The production of heavy fluids similar to crude oil is of interest from two points of view: the minimization and mitigation of waste streams and the production of alternative energy sources. Many different potential feed stocks for alternative crude oils have been explored, including swine manure, poultry parts and assembly line waste.

Indeed, we have examined some of these feed stocks with the ADC in the past. Recycling and reclaiming petroleum products for further use are topics under current development in industry. One product that could have further use as a feed stock for a crude oil is plastic produced as a waste stream from automotive factories. These solid waste plastics would normally be sent to a landfill. This resulting crude oil might then be utilized in a similar manner to traditionally drilled petroleum crude oils. In order to be a useful supplement to the traditional crude oil stream, it is important that the physical and chemical characteristics of resourced oil from waste plastics be evaluated. Thus, evaluating the volatility (as expressed by the ADC) of such a resourced crude oil would be of prime importance. We have chosen to discuss this fluid along with the discussion of the waste oils because many of the same issues of chemical decomposition are encountered.

MATERIALS AND METHODS

The n-hexane used as a solvent in this work was obtained from a commercial supplier and was analyzed by gas chromatography (30 m capillary column of 5%- phenyl-95%-dimethyl polysiloxane having a thickness of 1 µm, temperature program from 50-170°C at a heating rate of 5°C min^{-1}) with flame ionization detection and mass spectrometric detection (Huber *et al.*, 2009a). These analyses revealed the purity to be approximately 99% and the fluid was used without further purification.

The used lubricant oils were obtained as process streams from a commercial oil re-refining facility. These fluids were dark in color, with the viscosity of typical oils and were all single phase. They were examined by gas chromatography with flame ionization detection and mass spectrometric detection (GC-FID and GC-MS). The samples were analyzed on a 30 m capillary column of 5% phenyl-95%-dimethyl polysiloxane having a thickness of 1 µm. Initially, the temperature was maintained isothermally at 80°C for 2 minutes, followed by an 8°C min^{-1} ramp to 285°C and held at 285°C for 5 minutes. These analyses revealed the expected cluster of hydrocarbon compounds having the usual Gaussian-like distribution.

The used transformer and cutting oil was also obtained from a commercial oil re-refining facility. This fluid had a deep amber cast but was nevertheless transparent. Like the used lubricant oils, it was a single phase fluid, but was less viscous than the used lubricant oil. The re-sourced crude oil was obtained from a processor of waste or unused automotive plastics. This fluid was a dark gray-brown and was more viscous than the lubricant oil or the transformer oil. It, too, was single phase. A similar survey analysis was done on these fluids as was described above. These analyses revealed the pattern common to highly refined mineral oils, but with some minor polar constituents as well. These included fatty acid esters and ethers in the seven to eight carbons range. Neither silicones nor fluorinated compounds were found in the sample we examined.

Advanced Distillation Curves

For each experiment with the lubricant oils, 200 mL of a new or used oil was placed into the boiling flask of the distillation curve apparatus with a 200 mL pipette. The thermocouples were then inserted into the proper locations to monitor T_k, the temperature in the fluid and T_h, the temperature at the bottom of the take-off position in the distillation head. The uncertainty in the thermocouple measurements was 0.05°C. Enclosure heating was then commenced with a four-step program based upon a previously measured distillation curve. Volume measurements were made in a level-stabilized receiver.

For each experiment with the resourced waste plastics oil, an identical procedure was carried out with the exception of the scale of the experiment. For the resourced oil, 50 mL of the oil was distilled from a 100 mL round-bottomed flask. The temperature measurements were carried out in equivalent locations in the smaller apparatus and the uncertainties in the temperature measurements were the same. A smaller apparatus was used in the case of this fluid because of its significantly higher viscosity. In this case, the sample had to be introduced with a 50 mL syringe.

Each of the distillation curves was measured at ambient atmospheric pressure. The pressure was measured with an electronic barometer; the expanded total uncertainty (k = 2) in the pressure measurements was 0.003 kPa. Distillation temperature readings were corrected to what should be obtained at standard atmospheric pressure. This was done with the modified Sydney Young equation, in which the constant term was assigned a value of 0.000109 (Huber et al., 2009b; Andersen and Bruno, 2005; Widegren and Bruno, 2008a; 2008b; 2009; Bruno and Svoronos, 2004; 2005; Ott et al., 2008; Young, S., 1902; 1903; 1922; ASTM Standard D 2789-04b, 2005). The magnitude of the correction depends on the extent of departure from standard atmospheric pressure. The location of the laboratory in which the measurements reported herein were performed is approximately 1650 m above sea level (with an average atmospheric pressure of 83 kPa), resulting in a typical temperature correction of 7°C.

To provide the composition channel to accompany the temperature information on the distillation curves, sample aliquots were withdrawn for 10 selected distillate volume fractions. To accomplish this, aliquots of ~7 µL of emergent fluid were withdrawn from the sampling hammock in the receiver adapter with a blunttipped chromatographic syringe and added to a sealed autosampler vial containing a known mass of n-hexane solvent. A sample was withdrawn at the first drop of fluid from the condenser and then at each of nine additional predetermined volume fractions of distillate, for ten total sample aliquots.

The oil distillate fractions were evaluated by GCMS in scanning mode on the same column and with the same temperature program described earlier. Mass spectra were collected in scanning mode for each peak from 15-550 Relative Molecular Mass (RMM) units.

RESULTS AND DISCUSSION

Automotive Oils

Three commercially available (unused) automotive crankcase oils were measured with the ADC metrology. These measurements were done to provide a frame of reference for the volatility of lubricating oils and ultimately, used lubricating oils. Each fluid presented here was measured between three and six times. These oils are also called "petroleum" oils because they are produced in petroleum refineries. The distillation curves of the petroleum oils are presented in Fig. 1 as T_k, the temperatures measured in the boiling fluid, plotted against distillate volume fraction. These data are true thermodynamic state points that can be used to model each fluid with an equation of state. In this figure, the estimated uncertainty (with a coverage factor k = 2) in the temperatures is 0.3°C. The uncertainty in the volume measurement that is used to obtain the distillate volume fraction is 0.05 mL in each case. Each lubricating oil that we examined was a multi-grade automotive lubricant: petroleum based 5W- 30, 10W-40 and 20W-50 and a synthetic 5W-30. Synthetic oils are reported to be advantageous in cold starting situations because they lack the waxes and heavy paraffins found in petroleum based oils. Multigrade oils have viscosity improvers added to base petroleum oils to increase each oil's operating temperature range. A trend of higher distillation temperatures with higher viscosity rated oil is observed on Fig. 2. This trend is not surprising, since the higher viscosity oils generally have heavier, less volatile components. The somewhat flatter shape of the petroleum based 5W-30 oil (when compared to the other two petroleum based oils) indicates that this oil has a narrower composition range or contains compounds with similar boiling points.

We note that the petroleum based 5W-30 and the synthetic 5W-30 have similar distillation curves, but the distillation temperature of the synthetic oil leads the petroleum oil slightly. The maximum difference is 5°C at the 0.05 distillate volume fraction, gradually decreasing to less than 2°C at the 0.80 volume fraction. This indicates that the compounds used to formulate the synthetic 5W-30 are slightly less volatile than the compounds found in the petroleum 5W-30.

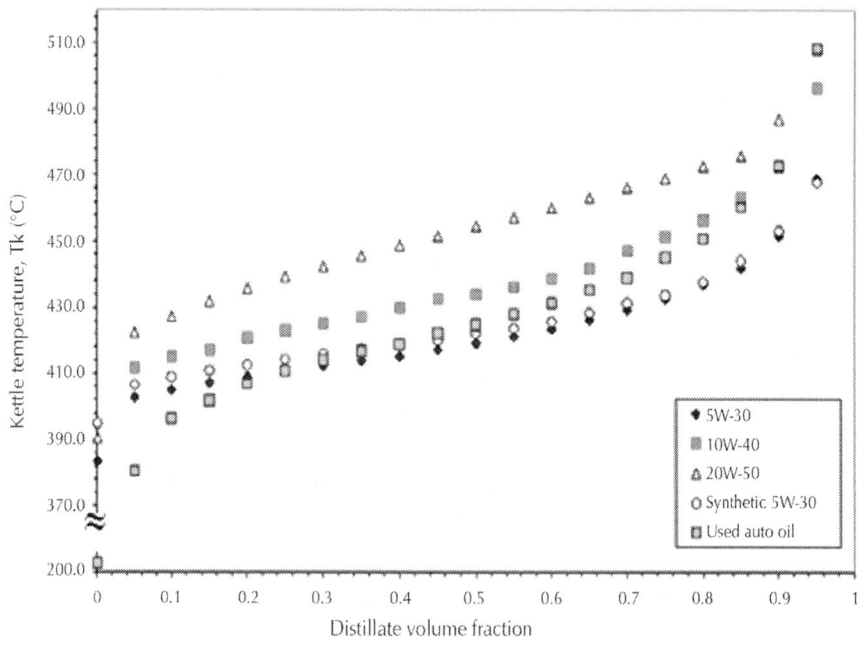

Figure 2: Distillation curves of three weights of virgin (unused) petroleum crankcase oil and a virgin synthetic crankcase oil. Also shown is the distillation curve for a stream of used crankcase oil. The error bars on the measurements are smaller than the symbols used.

The distillation temperatures of each of the petroleum automotive oils are all well above the upper temperature specification of 310°C for an atmospheric pressure distillation recommended by ASTM D-2892 to avoid "significant cracking". It is unlikely that the onset of severe cracking actually occurs until significantly higher temperatures, however, as discussed earlier. Hence, the application of an atmospheric distillation method at temperatures higher than 310°C is suitable with the ADC provided one uses the composition explicit data channel to monitor cracking. Moreover, the excellent repeatability (0.3°C) and the clear trends observed among the distillation curves of the petroleum derived lubricants suggest that an atmospheric distillation curve method can be a source of useful diagnostic information for such heavy fluids.

Also shown in Fig. 2 is the distillation curve of a sample of used automotive crankcase oils. The used automotive oil sample shown has not been treated or altered in any way; it is simply a composite sample presented to a re-refiner from multiple used oil collection centers. The first drop of the used automotive oil begins to distill at 207.7°C, which is more than 160°C lower than that of the unused oils. This suggests some gasoline contamination in the used oil. We did not find appreciable water in this sample (indeed, collection centers refuse to accept oil contaminated with a separate liquid water phase). The temperature profile of the used oil rapidly approaches that of the typical lubricating oil profile after this initial low temperature point. The early part of the distillation curve is near the curve for 5W-30. It then approaches the curve for 10W- 40 oil in the middle range, where it remains almost to the last fraction distilled. At the very end of the distillation curve, we note that the used oil curve climbs and merges with the distillation temperature of the 20W-50 oil at the end of the curve. This suggests that the used oil is made up of a mix of all three of these common oil weights, with a clear preponderance for the (most commonly used) 10W-40, although there is a small amount of residual fuel that distills early. This observation is consistent with expectations from a sample in an oil collection facility.

The composition-explicit data channel of the advanced distillation curve method can be used to probe the used automotive oil sample via GC-MS. Although a peak by peak analysis of these complex mixtures is possible and even desirable under some circumstances, some survey analyses can be performed to provide important information. For example, it is possible to evaluate the hydrocarbon moiety composition of each distillate fraction by use of a mass spectrometric classification method. For this we use a modification of the procedure suggested in ASTM Method D-2789 (ASTM Standard D 2789-04b, 2005). In this method, one uses mass spectrometry (or GC-MS) to characterize hydrocarbon samples into six types; it is convenient to think of D-2789 as presenting a hydrocarbon-type analysis of each distillate fraction. The six types or families are paraffins, monocycloparaffins, dicycloparaffins, alkylbenzenes (arenes or aromatics), indanes and tetralins (grouped as one classification) and naphthalenes. Although the method is specified only for application to low olefinic gasolines and it has significant limitations, it is of practical relevance to many complex fluid analyses and is often applied to gas turbine fuels, diesel fuels, rocket propellants and missile fuels. The uncertainty and the potential pitfalls of this method were treated earlier (Smith and Bruno, 2007a). Additionally, it is possible that not all of the components of the higher boiling fractions of distillate are sufficiently volatile to permit analysis by GC. In the present study, these higher boiling compounds are likely the heavy, greasy, hydrocarbons that comprise lubricant oils. Even in the absence of these heavy hydrocarbons, we would still be able to detect the families of lower boiling compounds that would be a primary concern for re-refining.

The solutions were prepared from 7 μL samples of distillate fraction that were dissolved in a known mass of solvent (n-hexane). This solvent was chosen because it causes no chromatographic interference with the sample constituents. For the hydrocarbon type analysis of the distillate fraction samples, 1 μL injections of the distillate diluted in hexane solvent were made into the GC-MS. Because of this consistent injection volume between samples, no corrections were needed for volume.

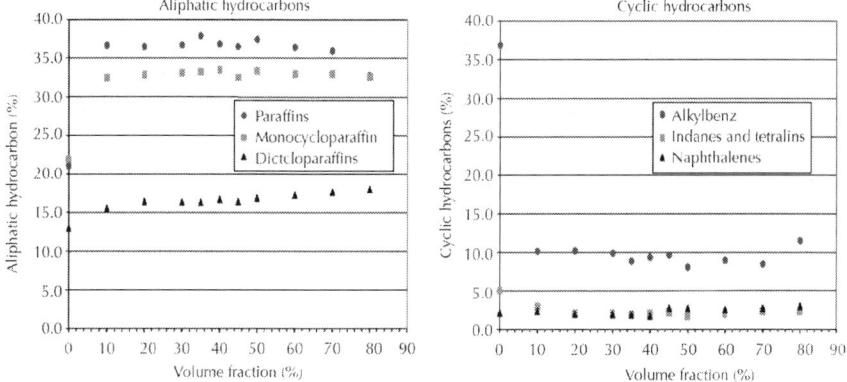

Figure 3: Hydrocarbon type survey analysis on the used automotive oil distillate fractions.

The results of the hydrocarbon type analysis are presented in Fig. 3. This Fig. 3 shows that the first drop of the used automotive oil distillate has significantly different chemical characteristics than those of the remaining 10 distillate fractions evaluated. Consistent with a separate, explicit investigation of the GC-MS peaks (discussed below), the first drop of distillate has a very high proportion of alkylbenzenes, containing nearly 37% (vol/vol), while the rest of the distillate fractions contain between 8 and 10% (vol/vol). Due to this high percentage of alkylbenzenes in the first fraction, the total percentage of paraffinic hydrocarbons in the first drop is significantly lower (approximately 22% for both paraffins and monocyloparaffins) than in the remaining distillate fractions. The dicycloparaffinic compounds are relatively unaffected; these percentages rise gently from 13-18% during the course of the distillation. The indanes/tetralins and naphthalenes are consistently low, with the indanes/tetralins showing a slightly higher percentage in the first drop. These major differences in the composition of the first drop of the used automotive oil are the cause of the very different distillation temperature of this drop compared to the rest of the distillation curve. We note that the results of this analysis are consistent with the suspicion of contamination of the used oil with residual fuel. The composition profile beyond the first drop is consistent with a lubrication oil, with a high content of paraffinics and monocyclic paraffinics and relatively low content of aromatics.

After this moiety-by-moiety characterization, which gives us a survey profile of each distillate fraction's composition, we can use the composition explicit data channel to explicitly investigate the composition of the first drop of the used automotive oil. We focus on this fraction explicitly for additional analysis: (1) because it is a departure from the remainder of the composition profile and (2) to confirm our preliminary conclusion or suspicion regarding its cause. For this analysis, we applied GC-MS, with a method developed and optimized for a light component in a heavier hydrocarbon liquid matrix (30 m capillary column of 5% phenyl-95%-dimethyl polysiloxane having a thickness of 1 µm, temperature was maintained isothermally at 50°C for 2 minutes, followed by a 8°C min^{-1} ramp to 285°C and held at 285°C for 5 min).

The results of this analysis show that the first drop of the used automotive oil does not contain a Gaussianlike "envelope" of peaks characteristic of a heavy hydrocarbon mixture, but instead contains several common components of gasoline (Smith and Bruno, 2007a), including shorter hydrocarbons, cyclic paraffins such as naphthalenes and indanes and some aromatics (primarily alkyl-substituted benzenes). While these data indicate that further treatment of this specific sample of used oil will be required before it is suitable for re-use, it also shows that for this particular sample of used oil, the residual fuel will only affect the very earliest distillate fractions.

Used and Resourced Heavy Oils

Although used automotive oil is the most abundant oil in the total oil waste stream of industrialized countries, there are also several other significant contributors. To include some of the other sources of used lubricant oil waste streams, we examined used cutting oil from machine shop operations, used transformer oil and a sample of commingled waste lubricant oils collected by a reprocessor in a nationwide facility. The latter sample of commingled oil typically contains a significant quantity of cutting fluid and coolant oil from large scale automated machine shop operations in industry, as well as smaller amounts of unclassified oils from smaller facilities. Additionally, we present the distillation curve of a resourced heavy crude oil, prepared from waste plastics from an automotive factory. Representative distillation curves, again presented as T_k and with the same uncertainties as given earlier, are presented in Fig. 4. As before, each curve was measured between 4 and 6 times.

Both the used commingled oil and cutting oil begin distilling at approximately 100°C, suggestive of the presence of dissolved or dispersed water in these samples. Indeed, the collected distillate in the receiver for the used cutting oil was clearly composed of two phase layers: the first 10 mL (which corresponds to 5 volume%) of the distillate was light yellow and transparent, which we have found to be characteristic of a primarily aqueous phase distilled from a mixed waste fluid stream (Ott *et al.*, 2008a). A dramatic rise in temperature was then observed and an abrupt change to a dark brown immiscible phase was observed to accumulate in the receiver. The presence of oil hydrocarbons in this phase was confirmed by GC-MS. These curves provide a valuable evaluation of these two oils, indicating that lower-boiling components would require re-refinement before these oils might be re-used.

The used commingled oil also began to distill around 100°C; unlike the cutting oil (which distilled around 100°C for only the first 5%, vol/vol), the first 25 volume percent of the distillate continues to boil at a much lower temperature (less than 203°C). Then, a temperature increase of 185°C occurs before the next 5 volume% distills. These characteristics of the distillation curve indicate that the used commingled oil has a wider composition distribution with a relatively high proportion of low-boiling components when compared to that of the other used lubricant oils.

The used transformer oil has a much flatter distillation curve than those of the other three used oils, with a total temperature change of only 85°C. Unlike the used automotive oil, which shows a temperature increase of greater than 125°C between the first drop of distillate and the 5 volume% fraction, the first boiling drop of the used transformer oil is within 2°C of the 5 volume% fraction. This indicates that the transformer oil does not contain lower boiling components from light contaminants, unlike the other two oils previously discussed. It follows that the used transformer oil likely needs less re-refinement than the other used oils before it might be suitable for re-use.

The difference in the three distillation curves discussed above can be probed with the composition explicit data channel and again by employing the moietyby- moiety analysis method described above for the automotive lubricants. As an example, the results of this analysis for the used transformer oil are shown in Fig 5. Consistent with the relatively flat distillation curve for this fluid, indicative of a narrow composition range, the moiety-by-moiety analysis presented in Fig. 5 shows that each distillate fraction has a similar composition.

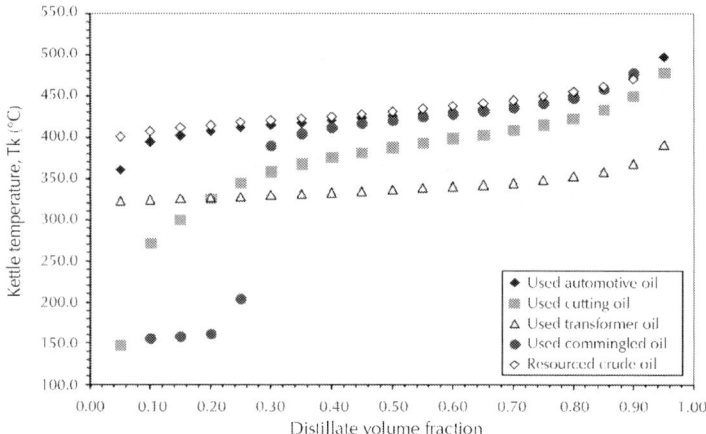

Figure 4: Distillation curves of the four used lubricant oils and the re-sourced crude oil. Each single curve shown for a given sample is representative of data collected for 3-6 distillation curves. The error bars on the temperature measurement are smaller than the symbols used.

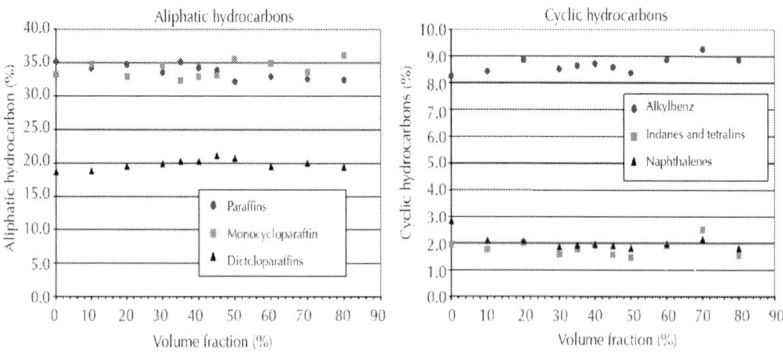

Figure 5: The hydrocarbon type survey analysis for the used transformer oil distillate fractions.

Paraffins and monocycloparaffins are the primary constituents of each distillate fraction, each moiety type being responsible for between 30 and 35% of the total composition. The dicycloparaffins are also a relatively large proportion of the total composition, varying between 19 and 21% throughout the course of the distillation. Each of the cyclic hydrocarbon moieties has a significantly smaller total percentage. Of the cyclic hydrocarbons, the alkylbenzenes have the highest overall percentages, between 8 and 10%. The indanes/tetralins and naphthalenes are consistently between 1 and 3% of the total composition. This narrow composition range is what would be expected for a highly refined fluid such as transformer oil. Additionally, the composition profile throughout the course of the distillation confirms the absence of the low-boiling species found in the other used lubricant oils.

Finally, we discuss the distillation curve of a sample of crude oil prepared from re-sourcing waste plastic from an automotive plant. This oil is intended to divert plastics from the landfill stream and to supplement petroleum crude oil reserves. Examination of the distillation curve of this resourced oil shows that the distillation curve has a relatively flat shape, indicative of a narrow compositional profile. No lowboiling components are indicated, as was found in some of the used oils. GC-MS investigation of the resourced oil before distillation showed that it was primarily composed of long-chain aliphatic hydrocarbon compounds, with naphthalenes and alkylbenzenes present as minor components. Examination of the composition of each distillate fraction shows the expected envelope of heavy hydrocarbon peaks, with the hydrocarbon envelope progressing to longer retention times as the distillation proceeds. However, each fraction also contains some peaks indicative of shorter, likely thermally cracked, hydrocarbons. Even with the observed cracking, the distillation curve is still useful because it determines that there are no lowboiling components in the resourced heavy crude oil.

CONCLUSIONS

The advanced distillation curve metrology was used to examine four unused automotive crankcase oils and four used lubricant oils, including a used automotive crankcase oil. The distillation curves of the three unused petroleum crankcase oils followed the expected trends based on the stated weights of the oils. A synthetic automotive crankcase oil was slightly less volatile than the petroleum oil of the same viscosity rating. Three out of the four used lubricant oils had distillation curves that indicated initial boiling at much lower temperatures than the remainder of the distillation curves, indicating that these three used oils contained lower boiling components that would require re-refining of the oil if the oil were to be re-used. The composition-explicit data channel was used to probe the composition of a low-boiling distillate fraction of the used automotive oil; several components of automotive gasoline were present. Moiety-by moiety analysis by a mass spectral method similar to ASTM D-2789 gave good insight into the composition of the oils as a function of distillate fraction. The resourced crude oil shows a relatively flat distillation curve, indicating the lack of low-boiling compounds. We conclude from this work that the advanced distillation curve metrology is a valuable diagnostic for new, used and resourced heavy oils. This method can be used to develop thermodynamic models for such complex fluids as mixed waste oils. These models are needed for the design and optimization of separation processes used in re-refining and reprocessing. Moreover, the method can be used as a QA/QC tool, since the chemical composition can be related to the thermal information that is measured.

ACKNOWLEDGEMENTS

LSO acknowledges a National Academy of Sciences/National Research Council (NAS/NRC) postdoctoral associateship at NIST; BLS acknowledges a Professional Research Experiences Fellowship (PREP) at NIST.

REFERENCES

1.   Andersen, P.C. and T.J. Bruno, 2005. Thermal decomposition kinetics of RP-1 rocket propellant. Ind. Eng. Chem. Res., 44: 1670-1676. DOI: 10.1021/ie048958g

2.   ASTM Standard D 2789-04b, 2005. Standard Test Method for Hydrocarbon Types in Low Olefinic Gasoline by Mass

Spectrometry. American Society for Testing and Materials, West Conshohocken, PA.

3.   ASTM Standard D 86-04b, 2004. Standard Test Method for Distillation of Petroleum Products at Atmospheric Pressure. ASTM International, West Conshohocken, PA.

4.   ASTM Standard D2887-02, 2004. Standard Test Method for Boiling Range Distribution of Petroleum Fractions by Gas Chromatography. ASTM International, West Conshohocken, PA.

5.   Bruno, T.J. and B.L. Smith, 2006a. Improvements in the measurement of distillation curves-part 2: Application to aerospace/aviation fuels RP-1 and S-8. Ind. Eng. Chem. Res., 45: 4381-4388. DOI: 10.1021/ie051394b

6.   Bruno, T.J. and B.L. Smith, 2006b. Enthalpy of combustion of fuels as a function of distillate cut: Application of an advanced distillation curve method. Energy Fuels, 20: 2109-2116. DOI: 10.1021/ef0602271

7.   Bruno, T.J. and P.D.N. Svoronos, 2004. CRC Handbook of Basic Tables for Chemical Analysis. 2nd Edn., Taylor and Francis CRC Press, Boca Raton, ISBN-10: 0849315735, pp: 688.

8.   Bruno, T.J. and P.D.N. Svoronos, 2005. CRC Handbook of Fundamental Spectroscopic Correlation Charts. 1st Edn., Taylor and Francis CRC Press, Boca Raton, ISBN-10: 0849332508, pp: 240.

9.   Bruno, T.J., 1994. Constituents and Physical properties of the C6+ fraction of natural gas. Topical Report, Gas Research Institute April-June Report No.: ISTIR-5212, GRI-94/0274; Order No. PB95136644.

10.   Bruno, T.J., 2006a. Improvements in the measurement of distillation curves-part 1: A composition-explicit approach. Ind. Eng. Chem. Res., 45: 4371-4380. DOI: 10.1021/ie051393j

11.   Bruno, T.J., 2006b. Method and apparatus for precision in-line sampling of distillate. Sep. Sci. Technol., 41: 309-314. ISSN: 0149-6395

12.   Bruno, T.J., A. Wolk and A. Naydich, 2009a. Stabilization of biodiesel fuel at elevated Temperature with hydrogen donors: Evaluation with the advanced distillation curve method. Energy Fuels, 23: 1015-1023. DOI: 10.1021/ef800740d

13. Bruno, T.J., A. Wolk and A. Naydich, 2009b. Composition-explicit distillation curves for mixtures of gasoline with four-carbon alcohols (butanols). Energy Fuels, 23: 2295-2306. DOI: 10.1021/ef801117c

14. Bruno, T.J., A. Wolk and A. Naydich, 2009c. Analysis of fuel ethanol plant liquor with the composition explicit distillation curve approach. Energy Fuels, 23: 3277-3284. DOI: 10.1021/ef900077t

15. Bruno, T.J., A. Wolk, A. Naydich and M.L. Huber, 2009d. Composition explicit distillation curves for mixtures of diesel fuel with dimethyl carbonate and diethyl carbonate. Energy Fuels, 23: 3989-3997. DOI: 10.1021/ef9002l5v

16. Cherepitsa, S.V., S.M. Bychkov, A.N. Kovalenko, A.L. Mazanik and N.M. Makoed et al., 2003. Determination of inspection parameters of diesel fuels. Chem. Technol. Fuels Oils, 39: 364-370. DOI: 10.1023/B:CAFO.0000011913.62536.d3

17. Emel'yanov, V.E., V.P. Grebenschikow, V.F. Golosova and G.N. Baranova, 1981. Influence of gasoline distillation curve on carburetor icing. Chem. Tech. Fuels Oils, 17: 619-621. DOI: 10.1007/BF00727813

18. Emel'yanov, V.E., V.P. Grebenshchikov, V.F Golosova and G.N. Baranova, 1982. Influence of gasoline distillation curve on carburetor icing. Khimiya Tekh. Top. Masel, 11: 22-23.

19. Hadler, A.B., L.S. Ott and T.J. Bruno, 2009. Study of azeotropic mixtures with the advanced distillation curve approach. Fluid Phase Equilibria, 281:49-59. DOI: 10.1016/J.FLUID.2009.04.001

20. Hallett, W.L.H. and M.A. Ricard, 1992. Calculations of the auto-ignition of liquid hydrocarbon mixtures as single droplets. Fuel, 71: 225-229. DOI: 10.1016/0016-2361(92)90013-E

21. Huber, M.L., B.L. Smith, L.S. Ott and T.J. Bruno, 2008a. Surrogate mixture model for the thermophysical properties of synthetic aviation fuel S-8: Explicit application of the advanced distillation curve. Energy Fuels, 22: 1104-1114. DOI: 10.1021/ef700562c

22. Huber, M.L., E. Lemmon, A. Kazakov, L.S. Ott and T.J. Bruno, 2009b. Model for the thermodynamic properties of a biodiesel fuel. Energy Fuels, 23: 3790-3797. DOI: 10.1021/ef900159g

23. Huber, M.L., E. Lemmon, L.S. Ott and T.J. Bruno, 2009a. Preliminary surrogate mixture models for rocket propellants RP-1 and RP-2. Energy Fuels, 23: 3083-3088. DOI: 10.1021/ef900216z

24. Huber, M.L., E.W. Lemmon, V. Diky, B.L. Smith and T.J. Bruno, 2008b. Chemically authentic surrogate mixture model for the thermophysical properties of a coal-derived-liquid fuel. Energy and Fuels, 22: 3249-3257. DOI: 10.1021/ef800314b

25. Karonis, D., E. Lois, F. Zannikos, A. Alexandridis and H. Sarimveis, 2003. A neural network approach for the correlation of exhaust emissions from a diesel engine with diesel fuel properties. Energy Fuels, 17: 1259-1265. DOI: 10.1021/ef020296p

26. Karonis, D., E. Lois, S. Stournas and F. Zannikos, 1998. Correlations of exhaust emissions from a diesel engine with diesel fuel properties. Energy Fuels, 12: 230-238. DOI: 10.1021/ef9700588

27. Kister, H.Z., 1988. Distillation Operation. McGraw-Hill, New York.

28. Kister, H.Z., 1991. Distillation Design. 1st Edn., McGraw-Hill, New York, ISBN: 10: 007034910X, pp: 729.

29. Lovestead, T.M. and T.J. Bruno, 2009a. Application of the advanced distillation curve method to aviation fuel avgas 100LL. Energy Fuels, 23: 2176-2183. DOI: 10.1021/ef8011189

30. Lovestead, T.M. and T.J. Bruno, 2009b. Comparison of the hypersonic vehicle fuel JP-7 to the rocket propellants RP-1 and RP-2 with the advanced distillation curve method. Energy Fuels, 23: 3637- 3644. DOI: 10.1021/ef900096q

31. Manovyan, A.K., D.A. Khachaturova and V.V. Lozin, 1983. Method for determination of distillation curve of heavy petroleum products. Chem. Tech. Fuels Oils, 19: 259-261. DOI: 10.1007/BF00726868

32. OilRe-refining.aspx, 2009. http://www.safetykleen. com/OilSolutions/OilRe-Refining/Pages/

33. Orbital Engine Company, 2003. Market barriers to the uptake of biofuels study: A testing based assessment to determine the impacts of a 20% ethanol gasoline fuel blend on the Australian passenger fleet.

34.   Ott, L.S. and T.J. Bruno, 2007a. Corrosivity of fluids as a function of distillate cut: Application of an advanced distillation curve method. Energy Fuels, 21: 2778 - 2784. DOI: 10.1021/ef070195x

35.   Ott, L.S. and T.J. Bruno, 2007b. Modifications to the copper strip corrosion test for the measurement of microscale samples. J. Sulfur Chem., 28: 493-504.

36.   Ott, L.S. and T.J. Bruno, 2008. Variability of biodiesel fuel and comparison to petroleum-derived diesel fuel: Application of a composition and enthalpy explicit distillation curve method. Energy Fuels, 22: 2861-2868. ISSN: 0887-0624

37.   Ott, L.S., A. Hadler and T.J. Bruno, 2008d. Variability of the rocket propellants RP-1, RP-2 and TS-5: Application of a composition- and enthalpy-explicit distillation curve method. Ind. Eng. Chem. Res., 47: 9225-9233. DOI: 10.1021/ie800988u

38.   Ott, L.S., B.L. Smith and T.J. Bruno, 2008. Experimental test of the Sydney Young equation for the presentation of distillation curves. J. Chem. Thermodynam., 40: 1352-1357. DOI: 10.1016/J. JCT.2008.05.009

39.   Ott, L.S., B.L. Smith and T.J. Bruno, 2008a. Advanced distillation curve measurements for corrosive fluids: Application to two crude oils. Fuel, 87: 3055-3064. DOI: 10.1016/J.FUEL.2008.04.032

40.   Ott, L.S., B.L. Smith and T.J. Bruno, 2008b. Advanced distillation curve measurement: Application to a bio-derived crude oil prepared from swine manure. Fuel, 87: 3379-3387. DOI: 10.1016/J.FUEL.2008.04.038

41.   Ott, L.S., B.L. Smith and T.J. Bruno, 2008c. Composition-explicit distillation curves of mixtures of diesel fuel with biomass-derived glycol ester oxygenates: A fuel design tool for decreased particulate emissions. Energy Fuels, 22: 2518- 2526. ISSN: 0887-0624

42.   Shin, Y.G., 1997. Simulation of volatility of commercial gasoline based on major hydrocarbon species. KSME Int. J., 11: 714-725. DOI: 10.1007/BF02946342

43.   Sjogren, M., H. Li, C. Banner, J. Rafter and R. Westerholm et al., 1996. Influence of physical and chemical characteristics of diesel fuels and exhaust emissions on biological effects of particle extracts: A multivariate statistical analysis of ten diesel fuels. Chem. Res. Toxicol., 9: 197-207. ISSN: 0893-228X

44. Smith, B.L. and T.J. Bruno, 2006. Advanced distillation curve measurement with a model predictive temperature controller. Int. J. Thermophys., 27: 1419-1434. DOI: 10.1007/s10765-006-0113-7

45. Smith, B.L. and T.J. Bruno, 2007a. Improvements in the measurement of distillation curves: Part 3- application to gasoline and gasoline + methanol mixtures. Ind. Eng. Chem. Res., 46: 297-309. DOI: 10.1021/ie060937u

46. Smith, B.L. and T.J. Bruno, 2007b. Improvements in the measurement of distillation curves: Part 4- application to the aviation turbine fuel Jet-A. Ind. Eng. Chem. Res., 46: 310-320. ISSN: 0888-5885

47. Smith, B.L. and T.J. Bruno, 2007c. Compositionexplicit distillation curves of aviation fuel JP-8 and a coal based jet fuel. Energy Fuels, 21: 2853-2862. DOI: 10.1021/ef070181r

48. Smith, B.L., L.S. Ott and T.J. Bruno, 2008a. Composition-explicit distillation curves of diesel fuel with glycol ether and glycol ester oxygenates: A design tool for decreased particulate emissions. Environ. Sci. Tech., 42: 7682-7689. DOI: 10.1021/es800067c

49. Smith, B.L., L.S. Ott and T.J. Bruno, 2008b. Composition-explicit distillation curves of commercial biodiesel fuels: Comparison of petroleum derived fuel with B20 and B100. Ind. Eng. Chem. Res., 47: 5832-5840. DOI: 10.1021/ie800367b

50. Smith, B.L. and T.J. Bruno, 2008. Application of a composition-explicit distillation curve metrology to mixtures of jet-A + synthetic fischer-tropsch S-8. J. Propul. Power, 24: 618-623. ISSN: 0748-4658

51. Widegren, J.A. and T.J. Bruno, 2008a. Thermal decomposition kinetics of the aviation fuel Jet-A. Ind. Eng. Chem. Res., 47: 4342-4348. DOI: 10.1021/ie8000666

52. Widegren, J.A. and T.J. Bruno, 2008b. Thermal decomposition of RP-1 and RP-2 and mixtures of RP-2 with stabilizing additives. Proceeding of the 4th Liquid Propulsion Subcommittee, JAANNAF.

53. Widegren, J.A. and T.J. Bruno, 2009. Thermal decomposition kinetics of propylcyclohexane. Ind. Eng. Chem. Res., 48: 654-659. DOI: 10.1021/ie8008988

54. You can make a difference by recycling your used motor oil, 2009. American Petroleum Institute. http://www.recycleoil.org/

55. Young, S., 1902. Correction of boiling points of liquids from observed to normal pressures. Proc. Chem. Soc., 81: 777.

56. Young, S., 1903. Fractional distillation. 1st Edn., Macmillan and Co., Ltd., London, pp: 284. Young, S., 1922. Distillation Principles and Processes. 1st Edn., Macmillan and Co., Ltd., London, pp: 509.

Chapter 9

Determination of Zinc-Based Additives in Lubricating Oils by Flow-Injection Analysis with Flame-AAS Detection Exploiting Injection with a Computer-Controlled Syringe

Gustavo Pignalosa, Moisés Knochen, and Noel Cabrera

Departamento "Estrella Campos," Facultad de Química, Universidad de la República, Av. Gral. Flores 2124, Casilla 1157, Montevideo 11800, Uruguay

ABSTRACT

A flow-injection system is proposed for the determination of metal-based additives in lubricating oils. The system, operating under computer control uses a motorised syringe for measuring and injecting the oil sample (200 µL) in a kerosene stream, where it is dispersed by means of a packed mixing reactor and carried to an atomic absorption spectrometer which is used as detector. Zinc was used as model analyte. Two different systems were evaluated, one for low concentrations (range 0–10 ppm) and the second capable of providing higher dilution rates for high concentrations (range 0.02%–0.2% w/w). The sampling frequency was about 30 samples/h. Calibration curves fitted a second-degree regression model ($r^2 = 0.996$). Commercial samples with high and low zinc levels were analysed by the proposed method and the results were compared with those obtained with the standard ASTM method. The t test for mean values showed no significant differences at the 95% confidence level. Precision (RSD %) was better than 5% (2% typical) for the high concentrations system. The carryover between successive injections was found to be negligible.

INTRODUCTION

Lubricating oils consist of a base of mineral or synthetic oil and several substances added in order to enhance different properties of the product [1]. Some of these additives are salts of organic acids and metals such as calcium, barium, magnesium, and zinc. Depending on the additive and the characteristics of the oil, metal concentrations range typically from 0.2 to 2 g/L. The concentration of the additives should be determined, either as a part of the quality control of the final products, or to provide information about the oil during its life cycle. This determination is carried out, according to standard methods from the American Society for Testing and Materials (ASTM), by means of either inductively coupled plasma optical-emission spectrometry (ICPOES) [2, 3] or flame atomic absorption spectrometry (FAAS) [4, 5]. These techniques are also used for the determination of wear metals in used lubricating oils, which is an activity carried out within predictive/proactive maintenance schemes of large engines and lubricated machinery.

From the analytical point of view, lubricating oils are a difficult matrix due to their high viscosity and hydrophobicity, which precludes direct introduction to standard nebulisers employed in ICP-OES and FAAS, as well as dilution with aqueous solvents. Thus, instrumental determinations usually require a dilution with organic solvent, for instance xylene, methyl-isobuthyl ketone, or kerosene. The samples are prepared by weight to avoid undesired volume uncertainties due to the viscosity of the oil. Organometallic metal standards are dissolved and diluted as necessary in the same solvent.

In order to avoid the need of organometallic standards, Wittmann [6] and Hon et al. [7] proposed the use of aqueous metal standards and an appropriate solvent chosen to allow the dilution of these standards. These procedures, as well as the standard methods mentioned before, share some drawbacks. All of them require intensive hand labour and large amounts of glassware are needed for performing the dilutions. This glassware should then be cleaned and prepared for future use.

The introduction of automation in the analytical laboratory [8, 9] permits several improvements, such as a reduction in the handling of samples and glassware, less solvent consumption, and reduced chemical wastes. However, the literature is rather scarce in automated methods for the analysis of oils. Flow-injection analysis (FIA) [10] has been the technique selected by several authors for this purpose, for the determination of the contents of either metal-based additives or wear metals. For instance, Granchi et al. [11] Proposed the use of an FIA-based system for the determination of metals in lubricating oils by ICP-OES, however the heavy task of sample preparation was carried out by a laboratory Robot, reducing the flow-injection system to the secondary task of sample introduction to the ICP spectrometer. A work by Borja et al. [12] suggested that FIA could be used for the determination of calcium-based additives in lubricating oils by means of emulsions and flame AAS detection. Online emulsification in an FIA system has also been proposed by Burguera et al. [13] for the automated determination of chromium in lubricating oils with electrothermal AAS detection.

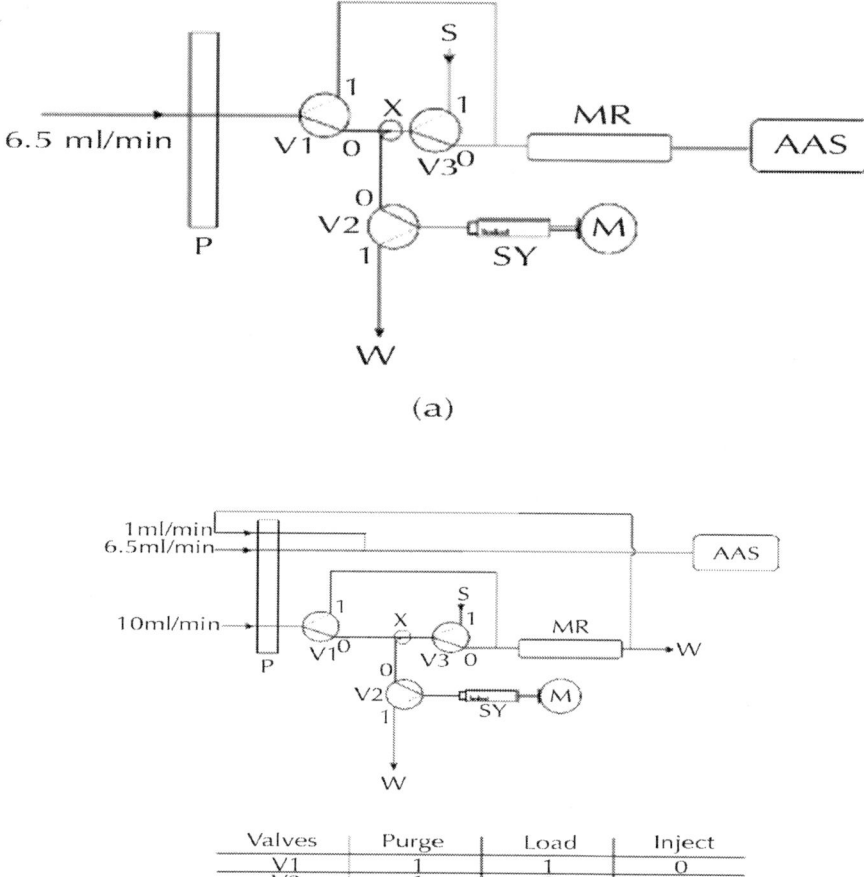

(a)

(b)

Valves	Purge	Load	Inject
V1	1	1	0
V2	1	0	0
V3	1	1	0

Figure 1: Schematic diagram of (a) the low-concentration system (LCS) and (b) the high-concentration system (HCS). P denotes peristaltic pump, V1, V2, V3 three-way solenoid valves ("0": default position; "1": energised position), X mixing device, S sample, SY syringe, M syringe motor, W waste, MR mixing reactor, and AAS atomic absorption spectrometer.

Pignalosa and Knochen [14] reported the use of a flow injection system for the determination of wear metals in lubricating oils, where injection of the sample was performed by a motorised syringe under computer control. This work demonstrated that syringe injection provides a robust means for accurately handling highly viscous oil samples.

On the other hand, metal concentration levels due to additives in the oil are usually much higher than those due to wear of the engine, thus the system must be capable of performing the necessary dilutions so that the final metal concentration in the sample bolus introduced to the nebuliser of the spectrometer is within a suitable range from the spectrometric point of view.

The objective of the present work is to explore the use of flow-injection analysis with syringe injection coupled to flame AAS as a way for automation of the determination of additives in lubricating oils in the routine quality control laboratory.

Zinc was chosen as a model analyte, considering that many lubricating oils contain high concentration of zincbased additives. On the other hand, there is a group of lubricating oils intended for use in special engines containing parts with silver alloys. These oils should not contain zinc as the presence of this element may damage the engine. For both reasons, the development of an automated method for the determination of zinc is necessary in quality control laboratories, and the same approach can be used for the determination of additives containing other metals.

EXPERIMENTAL

Apparatus

The flow-injection system (Figure 1) was developed around a lab-made motorised syringe, built from a Hewlett-Packard (Palo Alto, USA) 1mL gas-tight syringe, driven by a stepping motor and mechanism taken from a dismantled large diskette drive. The stepping motor employed had a resolution of $1.8°$/step, attaining a theoretical volumetric resolution of $2\ \mu L$ with a maximum dispensing volume of about $450\ \mu L$. On start-up, the "zero" position of the syringe was initialised with the help of an optocoupler used as optic position sensor, and the number of steps was counted from this original position.

Three 3-way 12-volt solenoid valves (model HP225T031, N Research, West Caldwell, USA) provided the necessary fluid switching. The carrier fluid was driven by a Dynamax RP-1 peristaltic pump (Rainin Instrument Co., Woburn, USA) fitted with Viton tubing. A lab-built mixing device (labelled "X" in Figure 1) and a packed mixing-reactor (labelled "MR" in the same figure), both already described elsewhere [14], were used to mix the streams of carrier and oil sample. The mixing device was machined in acrylic material and fitted with PTFE connectors (Omnifit, Cambridge, UK). The sample conduit was bored at an angle of 30°, respect to the carrier conduit, as previous experiments suggested that use of this angle produced less memory effects. The packed mixing reactor was constructed from a 6.0 cm length of 2.48mmID PTFE tubing and packed with small pieces of PTFE cut with a sharp blade from the same stock tubing. These pieces were basically 1mm thick tubing slices in turn cut into smaller pieces. This reactor was fitted with PTFE connectors (Omnifit).

Detection was performed by a Perkin-Elmer (Norwalk, USA) model 380 FAAS spectrometer with 10 cm burner and air-acetylene flame, fitted with a Photron (Narre Warren, Australia) zinc hollow cathode lamp and operated at the wavelength of 213.9 nm. In the system for high concentrations (see below), the burner was rotated as necessary to decrease the absorbance values to appropriate levels.

The operation of the system, data acquisition, and control were carried out by means of an 80MHz IBM compatible 80486-based personal computer fitted with a multipurpose data acquisition and control board (CIO-DAS- 08AOH, Computer Boards, Middleboro, USA) installed on the ISA bus. The 12 bit analogue-to-digital converter (ADC) in the card was used to capture analogue data from the spectrometer's recorder output, while several of the digital input and output (I/O) ports were used for logic control of the stepping motor and solenoid valves. On-board counters were used to provide appropriate timing when needed.

A special lab-built control system supplied the necessary power to the electronic devices, motor, and solenoid valves, and processed the signals to and from the multipurpose ADC board. The overall connection and signal flow schematics is depicted in Figure 2.

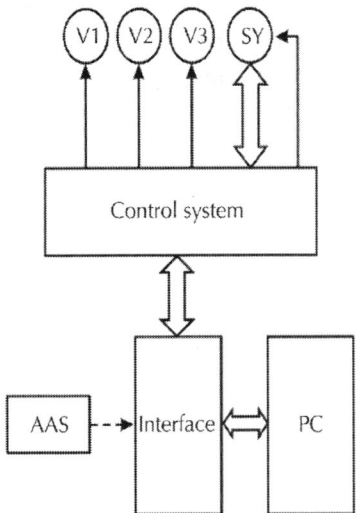

Figure 2: Schematics of connections and signals flow. V1, V2, V3 denote solenoid valves, SY motorised syringe, AAS atomic absorption spectrometer, PC personal computer, — power line, and - - - analog signal.

Software compiled in QuickBASIC 4.0 (Microsoft) was developed for the operation of the system and run under the MS-DOS 6.0 operating system. The program controlled the operation of the stepping motor and solenoid valves, triggered data acquisition via the ADC, and scaled the raw data transferred fromthe ADC to obtain absorbance values which were plotted on-screen as a function of time and stored on hard disk for further processing. Data was saved to disk as ASCII file and processed post run by means of a chromatography program (Peak Simple II, version 3.3, SRI Inc., Torrance, USA). A model 80550-20 strip-chart recorder (Cole-Parmer, VernonHills, USA) and a CR-6 recorder/integrator (Shimadzu, Kyoto, Japan) were also used for recording absorbance signals as a function of time.

Flow Systems

In order to handle samples containing both low and high concentrations of zinc, two different systems were developed. The first system (Figure

1a) was used in the determination of traces of zinc in oils supposed to be zinc-free. This system was used with Zn concentrations below 10 ppm and will be called "low-concentration system" (LCS). A second system (Figure 1b) was designed for the determination of high concentrations of zinc in those oils containing zinc-based additives ("high-concentration system" or HCS). In these oils, zinc concentrations as high as 0.15 %(w/w) can be found, making it necessary to implement an additional on-line dilution scheme. This system was used with Zn concentrations in the range 0.02%–0.2% (w/w). The operation of the two systems was similar, the main difference being the split-flow and confluence scheme used in the second system (Figure 1b) in order to produce further dilution with carrier of the sample bolus. This scheme corresponds to the classical cascade dilution system proposed by Whitman and Christian [15].

The motorised syringe (SY) could be connected either to waste or to the flow system via a three-way solenoid valve (V2). When sample had to be loaded, a special purge cycle was performed. For this purpose valves V1 and V3 were energised and carrier flowed to the AAS spectrometer by an alternate pathway. The syringe was driven backwards and sample was loaded into the syringe passing through V3 and the mixing device X, flushing the pathway with fresh sample and eliminating the rest of carrier or previous samples. When the appropriate volume of sample (300 μL) was reached, the syringe stopped, valve V3 was energised and the contents of the syringe expelled to waste. This purge cycle was carried out two times in order to minimise carryover effects. Afterwards, the syringe was loaded with 200 μL of sample, the three valves were turned off, and the syringe was driven forward, thus injecting the sample at a rate of 2.4mL/min into the flow system via the mixing device X, the solenoid valve V3, and the packed mixing reactor MR. From the output of this reactor, the dispersed sample bolus was either carried directly to the spectrometer's nebuliser (low-concentration system, Figure 1a) or submitted to further dilution with carrier via the splitting and confluence before sending it to the spectrometer (high-concentration system, Figure 1b).

Reagents, Standards, and Samples

The carrier was deodorised kerosene (ANCAP, Montevideo, Uruguay), which was used also as solvent in the reference (batch) method. This solvent was tested to be free of zinc. Conostan 5000 ppm zinc in oil standard (Conoco Specialty Products Inc., Ponca City, USA), diluted as necessary with BO-75 base oil (Conoco), was used for calibrations. Two groups of commercial lubricating oil samples, respectively with and without zinc-based additive, were used. Samples with additive were Rimula D 30, Helix Super 15W40, Tellus 46 (all from Shell), Superdiesel 40 (ANCAP), and GTS-99L1 and GTS-99L2 (both from Texaco, special noncommercial designations). Samples without added zinc were Ferrodiesel 597 (Repsol-YPF), Gascon Supreme plus MVI 40 (Lyondel), Viscodis 220 (ANCAP) and CAD-43 (Petrobras).

Methods

Standard solutions were prepared by exact dilution by weight of the Conostan standard with BO-75 base oil. Samples were introduced to the flow system without further processing. For purposes of validation, analytical results obtained with the proposed method were compared with those obtained by an ASTM batch flame AAS method [4]. In this method samples were diluted by weight with metal-free kerosene and measured by AAS with a nitrous oxide-acetylene flame.

VALIDATION

Linearity, accuracy, precision, detection limit (LOD), quantification limit (LOQ), carryover, stability, and sampling frequency were the figures of merit considered

Linearity

Linearity was assessed from calibration curves by means of the least-squares method (5 concentration levels).

Accuracy and Precision

Accuracy was evaluated by comparing the mean values of the results obtained by the proposed method with those obtained by the standard ASTM method. Student's *t* test for the mean values was used at the

95% confidence level. Precision was assessed from the results of these determinations.

Carryover

The possible existence of memory effects between successive injections was studied, in the low-concentration system by injecting a Viscodis 220 oil sample (containing about 8.7 ppm Zn) followed by injection of BO-75 base oil blank. For the high-concentration system, Superdiesel 40 oil (0.14% (w/w) Zn) and BO-75 were used for this purpose.

Stability

In order to assess the stability of the system during long runs, a sample of Superdiesel 40 oil containing about 0.14% (w/w) Zn was injected 30 times during a period of about 60 minutes.

RESULTS AND DISCUSSION

Injection Technique

Contemporary flow-injection analysis relies on the use of six port (or similar) injection valves for the introduction of the sample into the carrier stream. However in the early times of FIA, syringes were used for this purpose as attested by the first paper describing the technique [16].

When using injection valves, a sampling loop defines the volume of sample to be injected. In automated systems, a peristaltic pump is usually employed to fill this loop.

The difficulties associated with the handling of highly viscous samples such as lubricating oils are obvious. Peristaltic pumps, that are one of the usual pumping devices in automation, are of little value for handling this kind of samples, even for a simple task such as loading a sampling loop. In the first place, the flow rate is not independent from the viscosity (which in turn depends on the temperature and oil type). Besides, the pumping tubes have a high dead volume. When changing samples, it is extremely difficult to clean the pumping tubes and the sampling loop from the previous sample, if memory effects are to be avoided. This results in lengthy operations and low sample throughput.

In order to circumvent this problem, preliminary experiments were carried out trying to use the peristaltic pumps to generate negative pressure (i.e., sipping the sample through. The sampling loop), but this approach failed when applied to Lubricating oils.

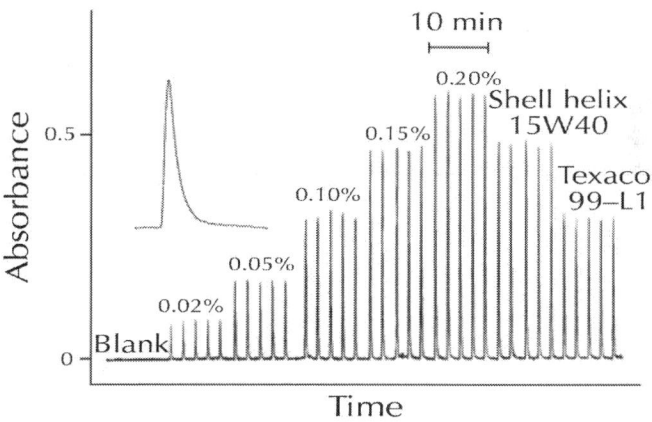

Figure 3: Recording of the signal corresponding to a calibration curve and two samples (concentrations in %(w/w)). Upper left angle: expanded trace of a typical peak.

In previous work [14], a motorised syringe has been successfully applied to the automated analysis of lubricating oils for wear metals. The piston motion demonstrated to be quite effective in displacing the oil out of the barrel, even with dirty samples, and good analytical precision could be achieved. Motorisation avoids the uncertainties associated with manual injection and is amenable to automation. Therefore, this approach was applied in the present work.

The use of a motorised syringe as injection device for oil samples deserves some comment. When the injection is performed via an injection valve fitted with a sampling loop, the sample is injected as a homogeneous "plug" which is subsequently dispersed in the carrier stream. When low viscosity samples are injected rapidly with a syringe as in early FIA work [16], the injected sample is supposed to displace an equal volume of carrier, resulting then in the injection of a sample "plug" which, once dispersed, will produce a mixing pattern and a peak shape similar to those obtained with loop injection. On the other hand, the slow injection of a viscous sample (such as an oil sample) produces a laminar flow pattern where both sample and carrier flow alongside with little mutual interaction unless some kind of mixing device is used to provide a vigorous mixing action.

In the present system, the injection had to be inherently slow because the viscosity of the sample would otherwise produce unacceptably high pressures inside the syringe barrel and other parts of the system. In this instance, the mixing action was provided by the packed mixing reactor ("MR" in Figure 1). Figure 3 shows the recording corresponding to a calibration curve and samples. The peak shapes obtained do not differ significantly from typical FIA peaks, suggesting that the radial mixing action of the mixing reactor is highly efficient. This was attributed to the random nature of the packing, which forces the flow into complex patterns with multiple direction changes.

Linearity

A second-degree model provided the best fit. Regression equation was

$$h = -4 \ 10^{-8}C^2 + 0.0004C \ (r^2 = 0.996) \quad\quad\quad (1)$$

Where h is the peak height (absorbance) and C is the concentration (ppm). Linear regression showed slightly worse fit ($r^2 = 0.993$), thus the second-degree model was chosen for calibration.

Accuracy and Precision

Analysis performed on the commercial samples showed no significant differences between the two methods at the 95% confidence level. Results can be seen in Table

Precision attained can also be seen in this table. Precision (RSD %) obtained with the HCS was always better than 5% and usually better than 2%. This was considered adequate for the analysis of this kind of samples. Precision is limited partly by the uncertainty in the injected volume. This in turn is affected by the resolution of the syringe.

Considering the mechanical resolution of the stepping motor used, the lab-made syringe employed had a volumetric resolution of 2 μL, which is 1% of the sample volume. It is predictable that better results could be obtained by using a motorised syringe with higher resolution.

Determinations carried out with the LCS exhibited higher dispersion, but this could be predicted given the low concentrations involved and hence the low absorbance values found.

Detection and Quantification Limits

For the LCS, LOD (3) was 0.08 ppm, and LOQ (10) was 0.28 ppm. For the HCS, LOD (3) was 33 ppm, and LOQ (10) was 110 ppm.

Carryover

In preliminary experiments, when injected repeatedly the BO-75 base oil blank, no signal was detected in either LCS or HCS systems. For the LCS, when injecting the BO-75 blank after the Viscodis 220 oil sample, no signal was detected.

When the same experiment was carried out in the HCS system with a Superdiesel 40 oil sample, the signal corresponding to the blank was less than 2.5% when compared with the signal of the Superdiesel 40 oil (Figure 4). This is the maximum extent of memory effect to be expected under the conditions established.

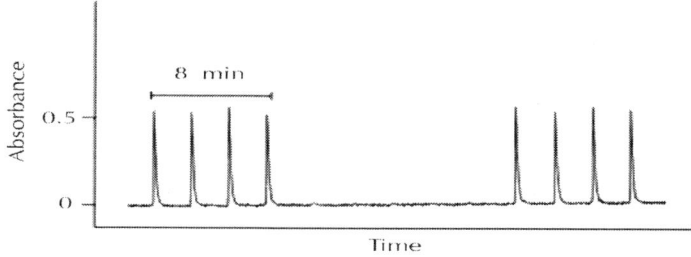

Figure 4: Signal recording corresponding to the test for memory effect. Four injections of Superdiesel 40 oil (Zn contents about 0.14% w/w), four injections of blank oil (BO-75), and four injections of Superdiesel 40.

Stability

Figure 5 shows the plot of the 60-minute run for the HCS. It can be observed that neither baseline nor response (peak height) varied significantly in that period, suggesting that the system is robust in long-term runs.

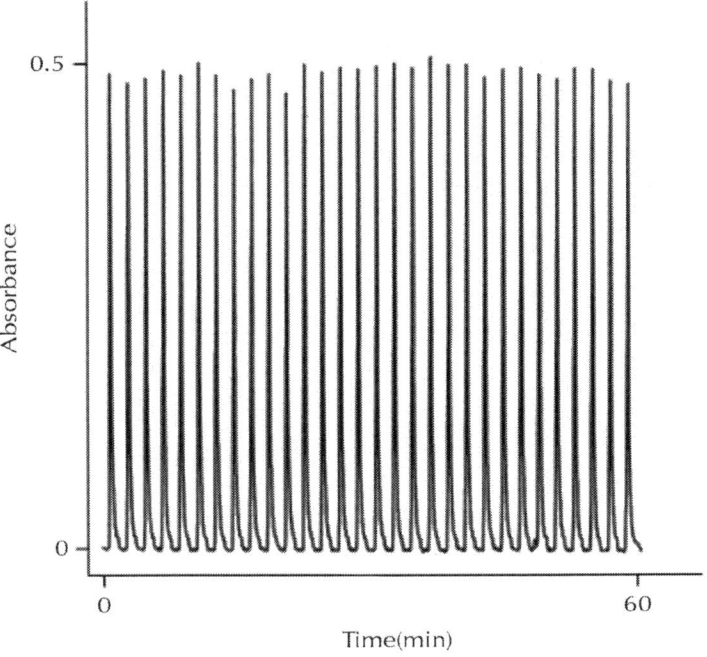

Figure 5: Signal recording corresponding to the stability test. A commercial sample (Superdiesel 40 oil) containing about 0.14% Zn was injected 30 times in a 1-hour period.

Sampling Frequency

Under the conditions established, the experimental sampling frequency was 30 samples per hour for the HCS, and around 40 samples per hour for the LCS.

Table 1: Comparison of the results obtained in the determination of zinc in commercial lubricating oil samples by means of the automated FIA system (proposed method) and the ASTM method (reference method). Mean values expressed either in ppm or %(w/w). n denotes number of repetitions, t_{exp} experimental value of Student's t statistic, t_{table} calculated value of Student's t statistic taken from the t distribution at the 95% confidence level, and (*) special noncommercial designations

Samples without Zn addition	Proposed method			Reference method					
Brand name	Mean (ppm)	RSD%	n	Mean (ppm)	RSD%	n	t_{ap}	t_{table}	Result
Repsol-YPF Ferrodiesel 597-1	0.9	12.5	5	0.9	5.0	5	0.33	2.31	✓
Repsol-YPF Ferrodiesel 597-2	0.6	14.3	4	0.6	7.5	5	0.79	2.37	✓
ANCAP Viscodis 220	8.7	4.1	5	8.3	1.4	5	1.96	2.31	✓
Petrobras CAD-43	2.5	3.8	4	2.5	3.6	5	0.09	2.37	✓
Lyondel Gascon MVI 40	1.1	20.2	5	1.1	1.6	5	0.10	2.31	✓
Samples with Zn addition	Proposed method			Reference method					
Brand name	Mean (%)	RSD%	n	Mean (%)	RSD%	n	T_{exp}	t_{table}	Result
Shell Helix 15W40	0.154	2.0	5	0.164	4.6	5	1.05	2.31	✓

Shell Tellus 46	0.036	4.7	5	0.035	2.5	5	0.15	2.31	✓
Shell Rimula D 30	0.080	1.0	5	0.078	3.9	5	0.52	2.31	✓
ANCAP Superdiesel 40	0.141	1.4	5	0.148	3.7	5	0.94	2.31	✓
Texaco GTS-99L1 (*)	0.097	2.7	5	0.113	6.0	5	1.85	2.31	✓
Texaco GTS-99L2 (*)	0.110	1.7	5	0.121	4.5	5	1.58	2.31	✓

CONCLUSIONS

The method of injecting the sample by means of a motorised syringe, associated with a packed mixing reactor has demonstrated to be a useful tool for the automation of the analysis of lubricating oil samples in a flow-injection system. These highly viscous samples can be successfully injected in a stream of organic solvent and dispersed into this solvent in a reproducible way.

When this system was applied to the automation of the determination of organometallic additives by AAS, satisfactory results were obtained, which compare favourably with those obtained with a reference method. It is concluded that the use of this kind of injection in flow-injection systems is a promising tool that extends the capabilities of flow-injection analysis to highly viscous samples.

ACKNOWLEDGMENTS

The authors thank Alexandra Sixto and Leticia Vazquez ´ for the enthusiastic cooperation with the AAS measurements. Partial financial support from UDELAR/CSIC and UNDP/PEDECIBA (Program URU/97/016) is gratefully acknowledged as well.

REFERENCES

1. T. V. Liston, "Engine lubricant additives, what they are and how they work," Lubrication Engineering, vol. 48, no. 5, pp. 389–397, 1992.

2. ASTM, "Standard Test Method for Determination of Additive Elements, Wear Metals, and Contaminants in Used Lubricating Oils and Determination of Selected Elements in Base Oils by Inductively Coupled Plasma Atomic Emission Spectrometry," in 1992 Book of ASTM Standards, Section 5, V. 5.03, American Society for Testing and Materials, Philadelphia, Pa, USA, 1992, Method D - 5185-91.

3. ASTM, "Standard Test Method for Determination of Additive Elements in Lubricating Oils by Inductively Coupled Plasma Atomic Emission Spectrometry," in 1992 Book of ASTM Standards,

Section 5, V. 5.03, American Society for Testing and Materials, Philadelphia, Pa, USA, 1992, Method D - 4951-89.

4. ASTM, "Standard Test Method for Determination of Additive Elements in Lubricating Oils by Inductively Coupled Plasma Atomic Emission Spectrometry," in 1992 Book of ASTM Standards, Section 5, V. 5.03, American Society for Testing and Materials, Philadelphia, Pa, USA, 1992, Method D - 4628-86.

5. M. de la Guardia and A. Salvador, "Flame atomic absorption determination of metals in lubricating oils: a critical review," Atomic Spectroscopy, vol. 5, pp. 150–155, 1984.

6. Z. Wittmann, "Direct determination of calcium, magnesium and zinc in lubricating oils and additives by atomicabsorption spectrometry using a mixed solvent system," Analyst, vol. 104, pp. 156–160, 1979.

7. P. K. Hon, O. W. Lau, and C. S. Mok, "Direct determination of metals in lubricating oils using atomic-absorption spectrometry and aqueous inorganic standards," Analyst, vol. 105, p. 919, 1980.

8. V. Cerda and G. Ramis, ` An Introduction to Laboratory Automation, Wiley, New York, USA, 1990.

9. P. B. Stockwell, Automatic Chemical Analysis, Taylor and Francis, London, UK, 2nd edition, 1996.

10. J. Ruzicka and E. Hansen, Flow Injection Analysis, Wiley, New York, USA, 2nd edition, 1989.

11. M.P. Granchi, J. A. Biggerstaff, L. J. Milliard, and P. Grey, "Use of a robot and flow injection for automated sample preparation and analysis of used oils by ICP emission spectrometry," Spectrochimica Acta B, vol. 42, pp. 169–180, 1987.

12. R. M. Borja, A. Salvador, M. de la Guardia, J. L. Burguera, and M. Burguera, Química Analítica, vol. 8, p. 241, 1989.

13. J. L. Burguera, R. A. de Salager, M. Burguera, et al., "On-line emulsification of lubricating oils in a flow-injection system for chromium determination by electrothermal atomic absorption spectrometry," Journal of Analytical Atomic Spectrometry, vol. 15, no. 5, pp. 549–555, 2000.

14. G. Pignalosa and M. Knochen, "Determination of wear metals in lubricating oils using flow injection AAS," Atomic Spectroscopy, vol. 22, pp. 250–257, 2001.

15. D. A. Whitman and G. D. Christian, "Cascade system for rapid on-line dilutions in flow-injection analysis," Talanta, vol. 36, no. 1-2, pp. 205–211, 1989.

16. J. Ruzicka and E. H. Hansen, "Flow injection analyses: Part i. A new concept of fast continuous flow analysis," Analytica Chimica Acta, vol. 78, no. 1, pp. 145–157, 1975.

10

Recycling of Waste Engine Oils Using a New Washing Agent

Ihsan Hamawand[1], Talal Yusaf[1], and Sardasht Rafat[2]

[1] Faculty of Engineering and Surveying, National Centre for Engineering in Agriculture, University of Southern Queensland, Toowoomba 4350, QLD, Australia.
[2] Faculty of Engineering and Science, University of Koya, Koya KOY45 AB64, Iraq.

ABSTRACT

This paper addresses recycling of waste engine oils treated using acetic acid. A recycling process was developed which eventually led to comparable results with some of the conventional methods. This gives the recycled oil the potential to be reused in cars' engines after adding the required additives. The advantage of using the acetic acid is that it does not react or only reacts slightly with base oils. The recycling process takes place at room temperature. It has been shown that base oils and oils' additives are slightly affected by the acetic acid. Upon adding 0.8 vol% of acetic acid to the used oil, two layers were separated, a transparent dark red colored oil and a black dark sludge at the bottom of the container. The base oils resulting from other recycling methods were compared to the results of this paper. The comparison showed that the recycled oil produced by acetic acid treatment is comparable to those recycled by the other conventional methods.

INTRODUCTION

Waste engine oil is a high pollutant material that requires responsible management. Waste engine oil may cause damage to the environment when dumped into the ground or into water streams including sewers. This may result in groundwater and soil contamination [1]. Recycling of such contaminated materials will be beneficial in reducing engine oil costs. In addition, it will have a significant positive impact on the environment [2–4]. The conventional methods of recycling of waste engine oil either requires a high cost technology such as vacuum distillation or the use of toxic materials such as sulfuric acid. These methods also produce contaminating by-products which have highly sulfur levels, especially in the Kurdistan region/Iraq. Lubricant oils have been used primarily for reducing friction between moving parts of various machinery or equipment, minimize material wear, and improve the efficiency of equipment /machinery and for fuel and energy savings. Access to lubricants is essential to any modern society and not only does lubrication reduce friction and wear by interposition of a thin liquid film between moving surfaces, but it also removes heat, keeps equipment clean, and prevents corrosion. One of its important applications includes gasoline and diesel engine oils [5]. Waste lubricating oil refers to the engine oil, transmission oil, hydraulic and cutting oils after use. It is also refers to the degradation of the fresh lubricating components that become contaminated by metals, ash, carbon residue, water, varnish, gums, and other contaminating materials, in addition to asphaltic compounds which result from the bearing surface of the engines [6]. These oils must be changed and removed from the automobile after a few thousand kilometers of driving because of stress from serious deterioration in service. The amount of lubricating oils that is collected annually in Europe and USA is very large, approximately 1.7 to 3.5 million tons. This large amount of waste engine oils has a significant impact on both economical and environmental aspects. They cost millions of dollars to manufacture and represent a high pollutant material when disposed of. If discharged into the land, water or even burnt as a low grade fuel, this may cause serious pollution problems because they release harmful metals and other pollutants into the environment [7].

A recommended solution for this issue is the recovery of the lubricating oil from the waste oil. Recycling processes using nontoxic and cost effective materials can be an optimum solution. Acid-clay has been used as a recycling method for used engine oil for a long time. This method has many disadvantages; it also produces large quantity of pollutants, is unable to treat modern multigrade oils and it is difficult to remove asphaltic impurities [8]. Solvent extraction has replaced acid treatment as the method of choice for improving the oxidative stability and viscosity/temperature characteristics of base oils. The solvent selectively dissolves the undesired aromatic components (the extract), leaving the desirable saturated components, especially alkanes, as a separate phase (the raffinate) [9]. In one study [10] a mixture of methyl ethyl ketone (MEK) and 2-propanol was used as an extracting material for recycling used engine oils. Although the oil resulting from this process is comparable to that produced by the acid-clay method, its cost is high. Expensive solvents and vacuum distillation are required to carry out this method [11,12]. Recently [13] propane was used as a solvent. Propane is capable of dissolving paraffinic or waxy material and intermediately dissolved oxygenated material. Asphaltenes which contain heavy condensed aromatic compounds and particulate matter are insoluble in the liquid propane. These properties make propane ideal for recycling the used engine oil, but there are many other issues that have to be considered. Propane is hazardous and flammable therefore this process is regarded as a hazardous process. Also, the extraction involves solvent losses, and highly skilled operating maintenance. In addition, extraction occurs at pressures higher than 10 atm and requires high pressure sealing systems which makes solvent extraction plants expensive to construct, operate and the method also produces remarkable amounts of hazardous by-products [14].

Membrane technology is another method for regeneration of used lubricating oils. In this method three types of polymer hollow fiber membranes [polyethersulphone (PES), polyvinylidene fluoride (PVDF), and polyacrylonitrile (PAN)] were used for recycling the used engine oils. The process is carried out at 40 ℃ and 0.1 MPa pressure. The process is a continuous operation as it removes metal particles and dusts from used engine oil and improves the recovered oils liquidity and flash point. Despite the above mentioned advantages, the expensive membranes may get damaged and fouled by large particles [15].

Vacuum distillation and hydrogenation are two other methods that can be used for recycling used engine oil [16]. The Kinetics Technology International (KTI) process is a combination of vacuum distillation and hydrofinishing. This method removes most of the contaminants from the waste oil. The process starts with atmospheric distillation to eliminate water and light hydrocarbons. This is then followed by vacuum distillation at a temperature of 250 ℃. The final stage is hydrogenation of the products to eliminate the sulphur, nitrogen and oxygenated compounds. This stage is also used to improve the color and odor of the oil. The product can be of quality standard (Gp.I) with a yield of approximately 82% and minimized polluting by-products. The disadvantage of this method is the high investment cost [17,18].

In this research glacial acetic acid was used for recycling used engine oils. The method provides a lower cost process in comparison with the conventional methods due to the low cost of the acid and the moderate conditions of the process. The recycled oil obtained by this method has been shown to have potential for reuse as an engine lubricant.

EXPERIMENTAL APPARATUS AND PROCEDURE

Experiment

For the purpose of this study a series of experiments have been carried out for two kinds of collected oils. The oils were collected from car oil change shops and from personal cars after 3000 to 3500 km in use. The type of base oil used in this study is Ravenol (VSi SAE 5W-40), manufactured in Werther, Germany. In order to confirm the ability of glacial acetic acid to separate the base oil from the contaminated substances many random experiments were carried out. The first experiment was carried out with the oils collected from the shops. The oil was heated to a temperature of over 250 ℮ for one hour, for the purpose of evaporating the water and the volatile substances in the used oil. This is to mimic the procedure used by the acid-clay method. The oil was then cooled to room temperature (16 ℮, winter) and then equal amounts of this oil (10 mL) were added to a number of beakers. Different quantities of glacial acetic acid were then added to the used oils in the beakers. The amounts of the acid added were 0.2, 0.4, 0.6, 0.8, 1.0, 1.2, 1.4, 1.6, 1.8, and 2 mL to each 10 mL sample. These samples then mixed at 600 rpm for half hour and heated on a hot plate stirrer (LabTech, ES35A-Pro) to a temperature of 25 ℮. The beakers were open to the atmosphere. Then the oils resulting from the stirring and heating were centrifuged (Sigma, 2-6E, Sigma Laborzentrifugen GmbH, Osterode am Harz, Germany) immediately for a quarter hour at a speed of 3000 rpm. No separation was observed after this process, besides the absence of change in the treated used oil color. This may be due to the heating which may result in evaporation of acetic acid. This may also be due to the low period of interaction between the acid with the used oil. The second set of experiments were done by increasing the mixing time to half an hour and later to one hour, followed immediately by centrifugation. For the second and third times no separation occurred and the absence of and color changes in the treated used oil was noted.

The third experiment was undertaken using covered beakers and the mixing process was performed for one hour at room temperature. The mixture was then left at room temperature for 24 hours before being placed in the centrifuge. Two layers were separated, a clear reddish lubricant oil layer and sludge at the bottom of the test tube. The lube oil separated from the sludge easily because the sludge was concentrated at the end of the test tubes. Both the lube oil and the sludge were weighed. In order to measure the weight of the sludge, the test tube was heated and then the sludge was removed from the tube using pieces of cotton. The cotton has been weighed in advance and the cotton with the sludge was dried in an oven (Memmert, UF110, Schwabach, Germany) at 50 ℃ for 24 hours. After drying, the weight of the sludge was confirmed to be 0.2 to 0.4 gm/10 gm sample of the used oil. The balance used in weighing is from Denver Instrument (Denver, S/SI-603, Denver, CO, USA).

The results indicated that the amount of sludge increased as the amount of acid added to the samples increased, up to a certain limit. Up to the 1.0 mL acid added/10 mL used oil the sludge collected was black in color, rigid and compacted in a small area in the bottom of the test tube after the centrifugation process. The sludge changed significantly after increasing the acid volume above 1.0 mL/10 mL used oil. The sludge became more like an emulsion and yellow in color, and it also occupied a quarter of the test tube. The experiments done with the first kind of the lubricant oil (the oil collected from the shops) were repeated for the second kind, personal cars lubricant oils of the Ravenol brand (VSi SAE 5W-40) collected after being in use for 3000 to 3500 km. The treatment procedure performed was similar to that applied for the oil collected from the shops. The experiments carried out using the used oil from personal cars gave better results regarding the clarity compared to the treated used oil collected from the oil-change shops. It worth noting that the oil-change shops in Koya city collect used oil from both compression ignition (CI) and spark ignition (SI) engines crankcases.

Experimental Procedures

Used cars lubricant oils have been collected from car oil-change shops in Koya city/Erbil-Iraq and personal cars after being in use for

3000 to 3500 km. These oils were collected in containers of 20 L each and then after mixing well by hand shaking for 15 min they were subdivided into smaller containers of 5 liters.

The recycling process started with removing the contamination from the used engine oil. Glacial acetic acid was mixed with the used engine oil using a ratio of 0.8 mL acetic acid to 10 mL used engine oil. This was followed by stirring in a closed container at room conditions (room temperature and pressure) for one hour. The mixture was left for 24 hours at room conditions and then underwent centrifugation for one hour to separate the base oil from the contaminants. The separated base oil was then mixed with kaolinite at a ratio of 1 mL oil to 4 g kaolinite. The kaolinite was added to remove the dark color and the smell which result from oxidation of some components in the oil. This was achieved by heating to a temperature slightly higher than 250 e, followed by centrifugation for 30 min. This process produced a yellow clear base oil and an amount of sludge very close to the initial weight of the kaolinite used plus 5 wt%. The sludge measurement followed the same procedure in Section 2.1. Figure 1 below illustrates the schematic diagram of the test rig used in carrying out the experiments.

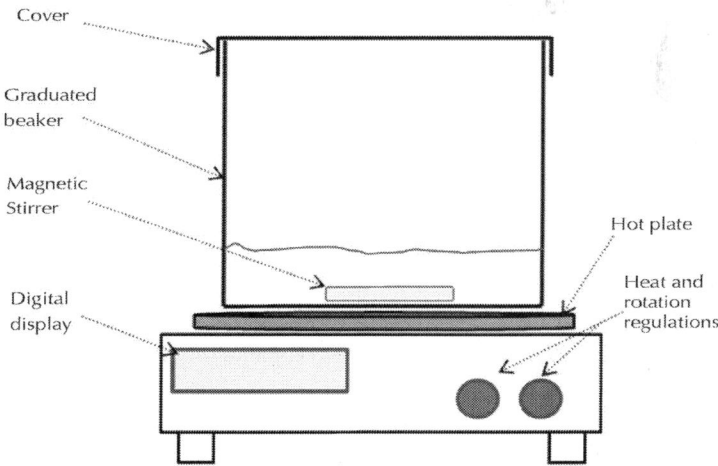

Figure 1: Schematic diagram of the test rig.

The second step in the recycling process was measuring the chemical and the physical properties of the resulting materials by performing oil analysis (OA). The materials that have undergone analysis are: the base engine oil produced by the acetic acid-clay method, base oil produced by local companies in Kurdistan region using the sulfuric acid-clay method, base oil resulting from vacuum distillation-clay treatment, base oils (Ravenol, VSi SAE 5W-40), used engine oil and the by-product (sludge).

Oil analyses (OA) is the laboratory analysis of the engine oils' physical and chemical properties, suspended contaminants and wear debris. Oil analysis is performed during routine preventive maintenance to provide meaningful and accurate information on lubricant and machine condition. By tracking oil analysis sample results over the life of a particular machine, trends can be established which can help eliminate costly repairs. Oil analysis can be divided into three categories:

Analysis of oil physical and chemical properties including tests such as: Flash Points by Cleveland Open Cup Tester ASTM D 92 [19], Pour Point ASTM D 97 [20], Viscosity Index ASTM D 2270 [21], Water and Sediment ASTM D 4007 [22], Rams bottom Carbon Residue ASTM D 524 [23], Total acid number (TAN) ASTM D 664 [24], Total Base Number (TBN) ASTM D 4739 [25], Refractive Index ASTM D 1218 [26], and Density (Specific Gravity) ASTM D 1298 [27].

Analysis of metallic contained by atomic absorption spectrometer (AAS).

Analysis of oxidative components by a Fourier Transform Infrared Spectrometer (Thermo Scientific, Thermo Mattson Nicolet 300-FTIR). The carbonyl groups have absorption peaks in the range of 1700 to 1750 cm^{-1}.

The metallic content analysis was performed by atomic absorption spectrometry using a fast sequential atomic absorption spectrometer (Varian, AA280 FS). Before the analysis the used engine oil sample was heated to 60 ℮ and stirred to ensure homogeneity of the sample, it was then mixed with 10 volumes of kerosene [28]. Sets of organometallic standards of metal (Zn, Cd, Cu, Fe, Ca, Pb, Sn, Mg and Mn) 4-cyclohexylbutyric acid salts were prepared and metal concentrations were determined by introducing the test solutions of engine oil samples into the flame of the atomic absorption spectrophotometer and recording the responses. Metal concentrations were determined from the calibration curve that was obtained from standard solutions. Standard solutions for all metals in engine oil samples were prepared according to Varian Techtron Pty LTD and ASTM D 4628-2 [29,30].

The specification of the apparatus used to carry out the experiments and the oil analysis are presented in Table 1.

Table 1: Specifications of the apparatus used in the experiments

Apparatus	Type	Model	Specifications
Scale	Denver	S/SI-603	Scale range: 0–600 g, readability: 0.001 g and accuracy: 0.002 g
Oven	Memmert	UF110	Forced air circulation, resolution of display for set point values 0.1 °C up to 99.9 °C, 0.5 °C from 100 °C to 300 °C, dimension (w ×h ×d): 560 ×480 ×400 mm.
Hot plate	LabTech	ES35A-Pro	Speed range: 0–1500 rpm, heating rate 6 K/min, heating temp. Range: ambient to 340 °C, dimension: 280 ×160 × 85 mm, display: analog scale.
Centrifuge	Sigma	2-6	Speed pre-selection up to 4000 rpm, low speed operation from 100 rpm, stainless steel bowl
Atomic Absorber spectrometer (AAS)	Varian	AA280 FS	Flame AA instrument, PC-controlled, double-beam Atomic Absorption spectrometer with Fast Sequential operation.
Fourier transform Infrared spectrometer (FT-IR)	Thermo Scientific	Thermo Mattson Nicolet 300-FTIR	FTIR machine equally handles solids and liquids

RESULTS AND DISCUSSION

Physical and Chemical Properties

Flash Point

The flash point of an engine oil is the lowest temperature to which the oil must be heated under specified conditions to give off sufficient vapor to form a mixture with air that can be ignited spontaneously by a specified flame. The flash point of engine oil is an indication of the oil's contamination. A substantially low flash point of an engine oil is a reliable indicator that the oil has become contaminated with volatile products such as gasoline. In the presence of 3.5% fuel or greater in used engine oils the flash point will potentially reduce to below 55 ℃. The flash point is also an aid in establishing the identity of a particular petroleum product. The flash point increases with increasing molecular mass of the oil. Oxidation would result in formation of volatile components which leads to decrease the flash point. [31]. Table 2 shows flash point values of different samples. The flash point of the base oil (Ravenol, VSi SAE 5W-40) is 232 ℃ because it is contains many different additives which contribute to improving its flash point. In contrast, the flash point of the measured used engine oil is 158 ℃. This decrease in flash point is a result of contamination with fuel and oxidation products [31]. On one hand, the table clearly shows that the flash point of the base oil is highly affected by adding sulfuric acid (185 ℃, row 9). On the other hand it is slightly affected by adding acetic acid (210 ℃, row 10) which indicates the destructive and complex effect of sulfuric acid even on the virgin oil. The flash point of the base oil produced using acetic acid treatment combined with vacuum distillation (220 ℃, row 4) is giving comparable results to that produced by professional companies using sulfuric acid (222 ℃, row 2) and the marketed base oil (232 ℃, row 1). The comparable results between the flash points of the two recycling methods are clearly showing the ability of glacial acetic acid to restore the used engine oil.

Table 2: Analysis of results of physical and chemical properties of different oil samples

No	Samples	Flash point °C	Pour point °C	Kinematic viscosity@ (40 °C)	Kinematic viscosity@ (100 °C)	Viscosity index	Refractive index	Specific gravity
1	Base engine oil (Ravenol, VSi SAE 5W-40)	232	-13	195.62	18	100.27	1.4886	0.8818
2	Marketed engine oil Recycled from used engine oil by professional recycling companies using acid (H_2SO_4) clay method and evaporation after adding the required additives	222	-10	169.5	16	97.07	1.4869	0.8828
3	Used engine oil (being in use for 2000–3000 km)	158	-5	136.6	13.5	89.11	1.4763	0.9261
4	Oil resulting from recycling the used engine oil by acetic acid (adding CH_3COOH + mixing + clay treatment + centrifugation) + vacuum distillation	220	-9.8	61.5	7.8	89.00	1.4833	0.8697
5	Oil resulting from recycling the used engine oil by acetic acid (adding CH_3COOH + mixing + clay treatment + centrifugation)	190	-9	80.25	8.5	82.47	1.4812	0.8768

6	Oil resulting from recycling the used engine oil by sulfuric acid (adding H_2SO_4 + mixing + clay treatment + centrifugation)	200	−9	72.5	9	84.45	1.4822	0.8707
7	Oil resulting from pretreatment of the used engine oil by acetic acid (adding CH_3COOH + mixing + centrifugation)	165	−8	130.5	13	91.84	1.4812	0.8838
8	Oil resulting from pretreatment of the used engine oil by sulfuric acid (adding H_2SO_4 + mixing + centrifugation)	180	−7	120	12	95.22	1.4785	0.8707
9	Oil resulting from mixing the base engine oil with acetic acid (adding CH_3COOH + mixing)	210	−11.8	182.7	15.6	89.18	1.4865	0.8717
10	Oil resulting from mixing the base engine oil with sulfuric acid (adding H_2SO_4 + mixing)	185	−9.5	175.6	7.8	89.30	1.4786	0.8707

Pour Point

The pour point of an engine oil is the lowest temperature at which the oil will remain in a flowing state. Most engine base oils contain waxes and paraffins that solidify at cold temperatures. Engine oils with high wax and paraffin content will have a higher pour point. Pour point is highly affected by an oil's viscosity, and engine oils with high viscosity are characterized by having high pour points. The pour point of an engine oil is an important variable, especially when starting the engine in cold weather. The oil must have the ability to flow into the oil pump and then be pumped to the various part of the engine, even at low temperatures [32]. Table 2 shows the pour point values of different samples. Acids are added to the base oil to examine their effect on the oil's composition. The pour point of the base oil was not notably affected by the addition of acetic acid. There was a small change in the pour point of around 1.2 ℃, but this value rose to 3.5 ℃ when sulfuric acid was added. It is also obvious that recycling the used oil with acetic acid combined with a vacuum distillation gave remarkable results (−9.8 ℃) and it is very comparable to the oil recycled using the acid-clay method (−10 ℃). The result of the acetic acid method is before adding the required additives to the oil. It is important to mention that the oil treated by acid-clay method before introducing the required additives only measures a pour point of −9 ℃ (row 6).

Kinematic Viscosities

Viscosity is a state function of temperature, pressure and density. There is an inverse relationship between viscosity and temperature, when the temperature of the engine oil decreases the viscosity increases and *vice versa*. Viscosity testing can indicate the presence of contamination in used engine oil. The oxidized and polymerized products dissolved and suspended in the oil may cause an increase of the oil viscosity, while decreases in the viscosity of engine oils indicate fuel contamination [33]. Oxidation of base oils during use in an engine environment produces corrosive oxidized products, deposits, and varnishes which lead to an increase in the viscosity [33]. Kinematic viscosity for different samples at two temperatures is shown in Table 2. The same viscosity test results have been predicted when acetic acid is added to the base oil at 40 and 100 °C. Kinematic viscosity at 40 °C was slightly affected by the addition of acetic acid to the base oil (182.7 cSt). The same result was predicted at 100 °C. The base oil lost only 13% of its original kinematic viscosity value. However, while the kinematic viscosity of the base oil was not greatly affected by the addition of sulfuric acid at 40 °C, it lost more than 50% of its value at 100 °C. These results show the destructive effect of sulfuric acid on the oil. It would appear that recycling the used oil with acetic acid (row 5) or sulfuric acid (row 6) is giving very comparable results at both temperatures (40 °C and 100 °C). The results of recycling with both methods reduced the kinematic viscosity of the used oil from 13.5 to 8.5 and 9 cSt, respectively. This means that the two methods are effective in removing the oxidized products, deposits, and varnishes from the used oil [33]. This may also mean removing the additives which have been added to enhance the viscosity.

Viscosity Index

Viscosity Index is strictly an empirical number and indicates the effect of change in temperature on viscosity. A high viscosity index indicates a small change in viscosity with temperature, which also means better protection of an engine that operates under vast temperature variations. Viscosity index improvers is among the common additives that improve the efficiency of the oil, however, engine oil with a high addition level of viscosity index improvers tends to degrade more rapidly. A high viscosity index is due to the absence of aromatic and volatile compounds. It also means good thermal stability and low temperature flow behaviors [34]. Table 2 shows very similar effects of both acids on the viscosity index when added to the base oil. The table shows that the viscosity index of the recycled used oil with acetic acid 89.00 (row 4) deviated slightly from that of the used oil 89.11 (row 3). This clearly indicates the low destructive effect acetic acid has on the viscosity index improver additives. It also seems that clay treatment contributes to a reduction in the viscosity index by approximately 10%, reduced from 91.84 (row 7) to 82.47 (row 5).

Refractive Index

Refractive index (RI) is the ratio of the light velocity in vacuum to the light velocity in substances at a specific temperature. The measurement of the refractive index is very simple, and requires small quantities of the samples. The refractive index can be used to provide valuable information about the composition of engine oils. Low values of refractive index indicate the presence of paraffin material while high values indicate the presence of aromatic compounds. It is also used to estimate other physical prosperities such as molecular mass [35]. Table 2 shows a base oil refractive index value of 1.4886. This is due to the presence of additives like polymers, polar organic compound, organic compounds, different metals, copolymers of olefins and hydrogenated diene styrene copolymers [35]. These components increase the molecular mass of the base oil and consequently its refractive index [35]. The results also indicate the low destructive action of the acetic acid, mixing the base oil with this acid reduces the RI by a small fraction of around 0.2%. In contrast, the addition of sulfuric acid reduces the RI by 0.7%. The same results are predicted after the recycling of the used engine oil with acetic acid and sulfuric acid.

Density (Specific Gravity)

Specific gravity is the ratio of the mass of volume of substance to the mass of the same volume of water and depends on two temperatures, at which the mass of the sample and the water are measured. Specific gravity is influenced by the chemical composition of the oil. An increase in the amount of aromatic compounds in the oil results in an increase in the specific gravity, while an increase in the saturated compounds results in a decrease in the specific gravity. An approximate correlation exists between the specific gravity, sulfur content, carbon residues, viscosity and nitrogen content [36]. Used engine oils specific gravity increases with the presence of increasing amounts of solids in the used engine oil. One percent of weight of solids in the sample can raise the specific gravity by 0.007 [36]. Used engine oil is contaminated with oxidized and condensed products rich in carbon. As shown in Table 2, base oil has a specific gravity of 0.8818, which is low compared to the used engine oil specific gravity of 0.9261. The high value of specific gravity of used engine oil is due to the presence of oxidation products, metals and contamination. Both acids have almost the same effect on the specific gravity when added to the base oil. The oil's specific gravity resulted from treatment by acetic acid combined with a vacuum distillation (0.8697) which is quite good compared to the virgin oil (0.8818), however this is before the introduction of the required additives to the recycled oil.

Water and Sediments

Water is generally referred to as a chemical contamination when suspended in engine oils. Water contamination of engine oil affects the oil quality, condition and wear of engines in service. The water content in engine oil is governed by the oil composition, physicochemical properties, production technology and conditions of storage and use. Water created in engine oil is a result of: absorbing moisture directly from the air (oil is hygroscopic), condensation (humid air entering oil compartments), heat exchanger (corroded or leaky heat exchangers), combustion (fuel combustion forms water which may enter the lubricant oil through worn rings), oxidation (chemical reaction) and neutralization (when alkalinity improvers neutralize acids formed during combustion), and free water entry (during oil changes). Water can prompt a host of chemical reactions such as hydrolysis of compounds and atomic species including oil additives base stock and suspended contaminants. In combination with oxygen, heat and metal catalyst, water is known to promote the oxidation and the formation of free radicals and peroxide compounds. Water attacks additives such as oxidation inhibitors, rust inhibitor, viscosity improver and the oil's base stock forming sludge. The water and sediment content of engine oil is significant because it can cause corrosion of equipment and problems in processing [37]. As shown in Table 3, base oil has water and sediment content values of zero due to the dehydration steps involved in producing base oils. In addition, introducing additives to the oil is minimizing the ratio of water and sediment to zero or near zero. It is obvious from the table that adding sulfuric acid to the virgin oil leads to different results to those seen when adding acetic acid. Water and sediments content of the base oil after adding the sulfuric acid is 0.6 mL/25 mL sample, compared to adding acetic acid <0.01 mL/25 mL sample. This clearly states the high reactivity of sulfuric acid with the materials that form the oil, one reaction being oxidation. The sulfuric acid-clay treatment shows a water content of <0.01 mL/25 mL sample because this method involves evaporation of the oil for more than one hour which helps in reducing the water content to this level. The same result as the sulfuric acid-clay method was achieved when adding acetic acid combined with vacuum distillation.

Carbon Residue

The amount of carbonaceous residue remaining after thermal decomposition of engine oil in a limited amount of air is also called coke or carbon forming tendency. The test for carbon residue can be used at the same time to evaluate the carbonaceous depositing characteristics of engine oils used in internal combustion engines. The carbon residue value of engine oil is regarded as indicative of the amount of carbonaceous deposits engine oil would form in the combustion chamber of an engine. It is now considered to be of doubtful significance due to the presence of additives in many oils. For example, an ash-forming detergent additive can increase the carbon residue value of engine oil yet will generally reduce its tendency to form deposits [38]. The carbon residue in base oil produced from raw petroleum materials is 0.55 wt%, which is 0.27 wt% lower than that produced by the recycling of used oil by the sulfuric acid-clay method, as shown in Table 3. This may be due to the complex reactions of the oil's components with sulfuric acid which may increase the sulfur content of the oil. A more precise relationship between carbon residue and hydrogen content, (H=C) atomic ratio, nitrogen and sulfur content have been shown to exist [38]. The recycling of the used oil by acetic acid combined with a vacuum distillation appears to provide better results. The carbon residue of this oil is slightly higher by 0.04 wt% than the base oil. As shown in the table, adding acetic acid to the base oil has increased the carbon residue by only 0.08 wt%. In contrast, adding sulfuric acid led to a greater increase in the carbon residue by 0.69 wt%. This clearly indicates the formation of components that have direct influence on the carbon residue of the oil.

Total Acid Number (TAN)

Total Acid number (TAN) is the weight (in milligrams) of potassium hydroxide required to neutralize one gram of the materials in the oil that will react with (KOH) under specific test conditions. The usual major components of such materials are organic acids, soaps of heavy metals. As engine oils are subjected to elevated temperatures, the process of oxidation occurs. Oxidation leads to the formation of organic acids in the engine oil. Total acid number (TAN) has been considered to be an important indicator for engine oil quality, specifically in terms of defining oxidation states. The presence of oxygen, in most engine oils environments, and hydrocarbons which make up the base oil lead to some reactions. This reaction may lead to the formation of carbonyl-containing products (primary oxidation products), subsequently these undergo further oxidation to produce carboxylic acids (secondary products) which results in an increase in the TAN value [39]. In addition, with time and elevated temperature, the oxidation products formed then polymerize leading to precipitation of sludge which decreases the efficiency of engine oil and causes excessive wear [39]. As shown in Table 3, the TAN of the used engine oil is measured to be 4.5 mg (KOH)/g (sample) which is much higher than the TAN of the base oils [0.02 mg (KOH)/g (sample)]. This is due to the presence of organic, inorganic, heavy metal salts, ammonia slots, resin, water and corrosive materials which result from the oxidation process that occurred at elevated temperatures in the engine [39]. The addition of sulfuric acid to the base oil has increased the TAN by 1 mg (KOH)/g (sample) compared to addition of acetic acid. It has been interpreted that the low acidity of the used oil when recycled with sulfuric acid is mostly due to the evaporation stage and clay processing rather than the acid action. The recycled oil by sulfuric acid without these two processes (row 8) is high in TAN around 2 mg (KOH)/g (sample).

Table 3: Analysis results of sediments and acidity of different samples

No.	Samples	Water and Sediment (mL)	Carbon residue (Wt %)	(TAN) mg KOH/ g(sample)	(TBN) mg KOH/ g(sample)
1	Base engine oil (Ravenol, VSi SAE 5W-40)	Zero	0.55	0.02	3.55
2	Marketed engine oil. Recycled from used engine oil by professional recycling companies using acid (H_2SO_4) clay method and steaming after adding the required additives	<0.02	0.82	0.14	2.8
3	Used engine oil (being in use for 2000–3000 km)	0.9	1.82	4.5	0.11
4	Oil resulting from recycling the used engine oil by acetic acid (adding CH_3COOH + mixing + clay treatment + centrifugation) + vacuum distillation	<0.01	0.59	0.15	-
5	Oil resulting from recycling the used engine oil by acetic acid (adding CH_3COOH + mixing + clay treatment + centrifugation)	<0.01	0.88	0.56	-
6	Oil resulting from recycling the used engine oil by sulfuric acid (adding H_2So_4 + mixing + clay treatment + centrifugation)	<0.01	0.99	0.7	-
7	Oil resulting from pretreatment of the used engine oil by acetic acid (adding CH_3COOH + mixing + centrifugation)	0.013	1.65	1.8	-

8	Oil resulting from pretreatment of the used engine oil by sulfuric acid (adding H_2SO_4 + mixing + centrifugation)	0.8	1.07	2	-
9	Oil resulting from mixing the base engine oil with acetic acid (adding CH_3COOH + mixing)	<0.01	0.62	1.5	2.5
10	Oil resulting from mixing the base engine oil with sulfuric acid (adding H_2SO_4 + mixing)	0.6	1.23	2.6	0.5

Total Base Number (TBN)

Internal combustion engine oils are formulated with a highly alkaline base additives package to neutralize the acidic products composition. The TBN is a measure of this package and it may be used as an indication for the engine oil's replacement time. This is because TBN depletes with time in service. Higher oil TBN values are more effective at neutralizing acids for longer periods of time. The rate of consumption of the additives is an indication of the projected service life of the oil [40]. As shown in Table 3 adding acetic acid to the base oil causes a small decline in the alkalinity of the oil from 3.55 to 2.5 mg KOH/g (sample). In contrast, adding sulfuric acid has reduced the alkalinity of the base oil by 96%. It is also obvious that the alkalinity is a result of the additives because the used oil's alkalinity is only 0.11 mg KOH/g (sample). TBN values for base oil (row 1) and that recycled using the sulfuric acid method (row 2) are 3.55 mg KOH/g (sample) and 2.8 mg KOH/g (sample), respectively. These values show higher TBN values than the TBN value of used engine oil [0.11 mg KOH/g (sample)]. This is due to the presence of a high alkaline additive package that was not depleted and has the ability to neutralize large quantity of acids, organic, inorganic bases including amino compounds and certain salts of heavy metals [40]. Used oil's TBN is low due to the depletion of the additive package as a result of the high temperature and the acid effect of water and oxidant product contamination. Base oil mixed with glacial acetic acid and concentrated 98% H_2SO_4 have TBN values 2.5 mg KOH/g (sample) and 0.5 mg KOH/g (sample), respectively. This clearly illustrates the lower effect of acetic acid on the oil's components.

Engine Oil's Metallic Content

Metals are regarded as heteroatoms found in engine oil mixtures. The amounts of metals are in range of a few hundred to thousands of ppm and their amounts increase with an increase in the boiling points or decrease in the API gravity of the engine oil. Engine oils' metallic constituents are associated with heavy compounds and they mainly appear in the residues. Base and base engine oils have very little metal content, which indicates their purity. Some metals present in virgin oils in high concentrations are in the form of various additives which improve the performance of the engine oil. Many others are introduced in to the oils after using due to depletion of various additives, engine bearings or bushings, and dilution of the engine oil with fuel containing metal additives [41]. Metals are found in used engine oil in two forms:

Metal Particulate Contamination

Metallic particulates enter the engine oil as a consequence of the breakdown of oil-wetted surfaces due to ineffective lubrication, mechanical working, abrasion erosion and/or corrosion. Metallic particles from deteriorating component surfaces are generally hard and increase the wear rate as their concentration in the oil increases.

Elemental (Metals)

Many oil constituents contain metallic elements that have been added to enhance the oil's efficiency. In general, metals in engine oils regarded as contaminants that should be removed completely in order to produce suitable base oil for producing new virgin oil [42]. Copper (Cu) is introduced to engine oils after use from bearings, wearing and valve guides. Engine oil coolers can also be contributing to copper content along with some oil additives [43]. The recycled oil with acetic acid combined with a vacuum distillation has a copper concentration of 0.4 ppm compared to zero ppm in oil recycled by sulfuric acid as shown in Table 4. This is because sulfuric acid is more reactive with copper than acetic acid and forms a precipitate. Magnesium is normally introduced into engine oil in an additive package. Magnesium is regarded as the most common wear metals in used engine oil and is present in virgin oil in the form of magnesium phenates and magnesium salicylates that behave as antioxidants at high temperatures [44]. Table 4 shows that the concentration of magnesium (Mg) in the base oil was affected more when sulfuric acid was added compared to acetic acid. This is again is due to the reactivity of sulfuric acid with the Mg, which may lead to precipitated compounds. Chromium presence in engine oil is normally associated with piston ring wear. High levels can be caused by dirt coming through the air intake or broken rings. Chromium may indicate excessive wear of chromed parts such as rings and liners [45]. As Table 4 is showing, the amount of chromium is very low, even in the used oil because this metal is persisting only in some parts of the engine. The recycling with acetic acid combined with a vacuum distillation seems to reduce the amount of this metal from 1.5 ppm to 0.2 ppm. The same result is shown in Table 4 in regards to tin. Tin is introduced into the oil after usd from piston wear, certain shaft types, bearings or bushings and valve guides [46]. Tin must be removed completely from base oils because it regarded as a contaminant. Lead (Pb) is associated with bearing wear, fuel source (leaded gasoline), and contamination due to the use of galvanized containers [47]. Lead concentration in virgin oils is zero because it is regarded as a base oil contaminant. Used engine oil has a lead concentration of 14.6 ppm as shown in Table 4. This is due to dilution of the engine oil by leaded gasoline that has been treated with tetraethyl lead as anti-knocking and engine part wear. The recycling with acetic acid combined with a vacuum distillation is very active in removing this metal. It helped to reduce the used oil content of this metal from 14.6 to 0.4 ppm. The most common wear metal in a car's engine that is introduced into the engine oil after a period of use is iron. Iron comes from many various places in the engine such as liners, camshafts and crank shaft, pistons, gears, rings, and oil pump. Iron concentration in engine oil depends on the bearing conditions inside the engine. If a bearing fails, iron concentrations in used engine oil increases. In the engine, the wear rises at a faster rate during the starting of the engine [48]. Base oils must be free from iron completely because it participates in producing oxidation products due to different chemical reactions. It is also clear that recycling with acetic acid combined with a vacuum distillation is very efficient in completely removing the metal. Zinc is introduced to base oil in the form of additives package as anti-oxidant, corrosion inhibitor, anti-wear, detergent and extreme pressure tolerance. Zinc is introduced in to base oil as additives, such as [49]:

- Zinc diethyldithiophosphate (ZDDP), which functions as an oxidation inhibitor that increases the oxidation resistance of the oil.

- Zinc dithiophosphates, this is not only acts as an anti-oxidant, but also acts as a wear inhibitor and protects the engine metals against corrosion.

- Zinc dialkyldithiocarbamates, this compound is mainly used as anti-oxidants but it is also has extreme pressure activity.

Zinc concentration in the base oil is 1,200 ppm, as shown in Table 4. This is added to the base oils as part of multi-functional additives for improving the oils' performance. The used engine oil content of zinc is 1280 ppm. The increase in the zinc concentration in used engine oil results from wearing of the galvanized piping. Adding sulfuric acid to the base oil leads to the reduction in zinc content by 28% compared to 11% with acetic acid. Sulfuric acid reacts with zinc to produce precipitated compounds. Manganese (Mn) is introduced from wear of cylinder liners, valves, and shafts [50]. As shown in Table 4 only a small amount of manganese is present in the used engine oil. Recycling with acetic acid combined with vacuum distillation is very effective in complete removal of this metal. Cadmium (Cd) is introduced in the engine oil as a contaminant during use. Base oils are free from cadmium [51]. It is clear from Table 4 that cadmium only has a count of 1 ppm in the used oil and it can be completely removed by acetic acid treatment.

Table 4: Analysis results of metal content of different samples

No.	Samples	Cu (ppm)	Mg (ppm)	Cr (ppm)	Sn (ppm)	Pb (ppm)	Fe (ppm)	Zn (ppm)	Mn (ppm)	Cd (ppm)
1	Base engine oil (Ravenol, VSi SAE 5W-40)	0	72	0	0	0	0	1200	0	0
2	Marketed engine oil. Recycled from used engine oil by professional recycling companies using acid (H_2SO_4) clay method after adding the required additives	0	68	0	0	0	0	1050	0	0
3	Used engine oil (being in use for 2000–3000 km)	4.6	81	1.5	1.6	14.6	72	1280	1.5	1
4	Oil resulting from recycling the used engine oil by acetic acid (adding CH_3COOH + mixing + clay treatment + centrifugation) + vacuum distillation	0.4	0.8	0.2	0.2	0.4	0	41	0	0
5	Oil resulting from recycling the used engine oil by acetic acid (adding CH_3COOH + mixing + clay treatment + centrifugation)	1.6	3.2	0.7	0.7	3.2	3.2	81.6	0.5	0.4

#										
6	Oil resulting from recycling the used engine oil by sulfuric acid (adding H_2SO_4 + mixing + clay treatment + centrifugation)	0.9	2	0.4	0.6	4.5	1.2	54	0.2	0.1
7	Oil resulting from pretreatment of the used engine oil by acetic acid (adding CH_3COOH + mixing + centrifugation)	3.56	65	0.8	0.9	11.2	34	780	0.8	0.45
8	Oil resulting from pretreatment of the used engine oil by sulfuric acid (adding H_2SO_4 + mixing + centrifugation)	2.1	45	0.5	0.6	5.4	24.4	613	0.4	0.2
9	Oil resulting from mixing the base engine oil with acetic acid (adding CH_3COOH + mixing)	0	65	0	0	0	0	1076	0	0
10	Oil resulting from mixing the base engine oil with sulfuric acid (adding H_2SO_4 + mixing)	0	57	0	0	0	0	865	0	0

Atomic Absorption Spectrometry (AAS) was used to measure the metals traces in the oil. The detection limits of the metals presented in Table 4 are shown in Table 5. It is obvious that the metals traces measured are far above the detection limit of the Atomic Absorption Spectrometry (AAS) used in the analysis.

Table 5: Detection limits of the AAS

Element	Symbol	Detection Limit (ppm)
Copper	Cu	0.003
Magnesium	Mg	0.0003
Chromium	Cr	0.006
Tin	Sn	0.1
Lead	Pb	0.01
Iron	Fe	0.006
Zinc	Zn	0.001
Manganese	Mn	0.002
Cadmium	Cd	0.002

Sludge Analysis

A secondary oxidation phase occurs at high temperatures where the viscosity of the bulk medium increases as a result of the polycondensation of the difunctional oxygenated products formed in the primary oxidation phase such as carboxylic acids. Further polycondensation and polymerization reactions of these high molecular weight intermediates form products which are no longer soluble in the hydrocarbon. The resulting precipitate is called sludge. Under thin-film oxidation conditions, as in the case of a lubricant film on a metal surface, varnish-like deposits are formed. The polycondensation reactions lead to high molecular weight intermediates (sludge precursors) [52].

The sludge produced has been analyzed for metals content. The analysis was done for the sludge collected from three resources; sludge resulting from centrifugation of the used oil directly before carrying out any further treatment (column three), after treatment with sulfuric acid-clay and after treatment with acetic acid-clay. The sludge collected from direct centrifugation was very low in quantity due to failure of centrifugation alone to precipitate contaminant in the used oil. As shown in Table 6, the sludge resulting from acetic acid-clay treatment is richer in metals than the one treated with the sulfuric acid-clay method. This high content of metals explains the lower losses in the sludge weight (~70%) when heated to 800 ℮ for three hours compared to sulfuric acid-clay method (~95%). This confirms the ability of acetic acid to precipitate the metal content in the used oil by 25% more than the sulfuric acid method.

Table 6: Analysis results of the sludge

Samples	Sludge Analysis		
	Sludge resulted from acidclay method + centrifugation	**Sludge resulted from acetic acidclay method + centrifugation**	**Sludge resulted from used oil**
Losing in ignition at 800 °C for 3 hours %	94.714	69.9417	95.448
Na	0.170	0.79	0.223
Cu	0.28	Nil	Nil
Pb	1.11	1.17	0.0736
Ag	Nil	Nil	Nil
Ni	0.058	0.64	Nil
Mg	0.28	3.74	Nil
V	Nil	Ni	Nil
Mo	1.33	1.91	Nil
Al	Ni	0.44	Nil
Sn	Nil	Nil	Nil
Ti	Nil	Nil	Nil
Ca	1.31	2.23	1.28
Ba	Nil	Nil	Nil
Cr	0.811	9.3	0.9655
SiO2	Nil	8.2	Nil
Total	99.5674	99.5217	98.0111

Color

Different categories of the engine oil were pictured to show the color changes that the oil went through during the process. As shown in Figure 2 and Table 7, the base oil (column 9) has changed color and become highly opaque (column 6) due to the vigorous reaction with sulfuric acid. In contrast, the base oil did not show any change in color when mixed with acetic acid (column 7) due to its low reactivity of acetic acid with the base oil. The acetic acid has shown a high reactivity with the contaminants in the used oil, and columns 5 and 8 show the base oil results from treatment with sulfuric acid and acetic acid, respectively. It is obvious that the base oil resulting from acetic acid treatment is brighter and has a clear yellowish color.

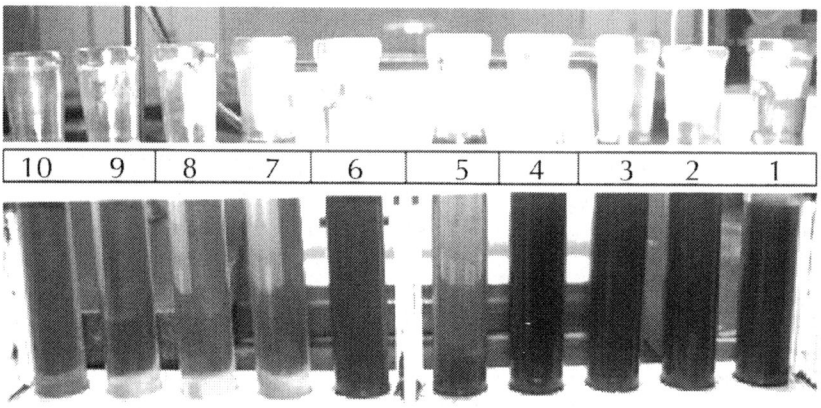

Figure 2: Pictures of the engine oil samples as described in Table 7.

Table 7: Description of the engine oil samples in Figure 2

No.	Description
1	Used engine oil
2	Used engine oil after centrifugation only
3	Used engine oil treated with sulfuric acid after mixing for 1 hour
4	Used engine oil treated with glacial acetic acid after mixing for 1 hour
5	Base oil produced by using sulfuric acid
6	Base oil (Ravinol type) + concentrated H_2SO_4 + mixing for 1 hour
7	Base oil (Ravinol type) + glacial acetic acid + mixing for 1 hour
8	Base oil produced by using glacial acetic acid
9	Base oil (Ravinol type)
10	Base oil produced by vacuum distillation and clay treatment

Engine Oil Oxidation Products Analysis

Oxidation of engine oil inside the engine is related to the availability of oxygen and in-cylinder pressure and temperature [52]. It can be divided into two types: oxidation at low and high temperatures. Oxidation of engine oil at low temperatures leads to alkylhydroperoxides ROOH, dialkylperoxides ROOR, alcohols ROH, aldehydes RCHO and ketones RR'C=O. In addition, cleavage of a dihydroperoxide leads to diketones $RCO(CH_2)_xCOR'$, ketoaldehydes $RCO(CH_2)_xCHO$, and hydroxyketones $RCH(OH)-(CH_2)_xCOR'$ [53]. At high temperatures (>120 ℃) the engine oil oxidation process can be divided into a primary and a secondary oxidation phase. In the primary oxidation phase the initiation and propagation of the radical chain reaction are the same as discussed under low-temperature conditions, but selectivity is reduced and reaction rates increased. At high temperatures the cleavage of hydroperoxides plays the most important role. Carboxylic acids (RCOOH) form, which represents one of the principal products under these oxidation conditions. In a subsequent step they can react with alcohols R'OH to form esters (RCOOR'). The termination reaction proceeds through primary and secondary peroxy radicals, but at temperatures above 120 ℃ these peroxy radicals also interact in a non-terminating way to give primary and secondary alkoxy radicals [54]. The secondary oxidation phase happens at higher temperatures where the viscosity of the bulk medium increases as a result of the polycondensation of the difunctional oxygenated products formed in the primary oxidation phase. Further polycondensation and polymerization reactions of these high molecular weight intermediates lead to form sludge [53]. Reaction oxidation compounds in oil samples determined qualitatively by obtaining their IR spectra in a Fourier Transform Infrared Spectrometer (Thermo Scientific, Thermo Mattson Nicolet 300-FTIR). A spectral band at 1700–1750 cm^{-1} indicated the presence of oxidation compounds, because of the fact that (C=O) bond strongly absorbs at this frequencies [55].

The IR results of the base oil as shown in Figure 3 and Table 8 showed no peaks of oxidized products which confirm its virginity. IR spectra for used engine oil show a medium peak at 1704.61 cm^{-1} that indicates the presence of oxidation products in the used engine oil sample, as shown in Figure 4 and Table 9.

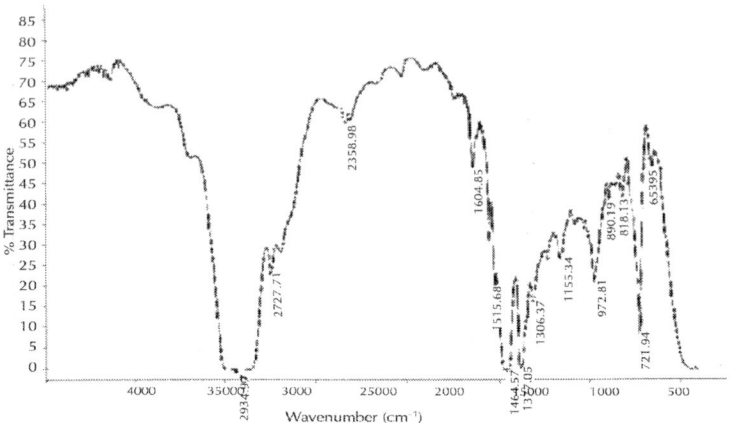

Figure 3: Results of analysis of base oil (Ravenol, VSi SAE 5W-40).

Table 8: Results of analysis of base oil (Ravenol, VSi SAE 5W-40)

Frequency cm⁻¹	Type of vibration	Bond	Functional group
653.95	Stretch	C-X	Alkyl halide
721.94	Bending in plain (roking)	C-H	Alkenes
818.13	Out of plan bending	C-H	Alkenes
890.19	Out of plan bending	C-H	Alkenes
972.81	Out of plan bending	C-H	Aromatic
1155.34	Wagging	CH2-X	Alkyl halide
1306.37	Wagging	CH2-X	Alkyl halide
1377.05	Out of plan bending	C-H	Alkans
1464.57	Bending in plain (scissoring)	C-H	Alkans
1515.68	Stretch	C=C	Aromatic
1604.85	Stretch	C=C	Aromatic
2358.98	Stretch	C ≡ C	Alkynes
2727.71	Stretch	C ≡ C	Alkynes
2934.97	Stretch	C-H	Alkans

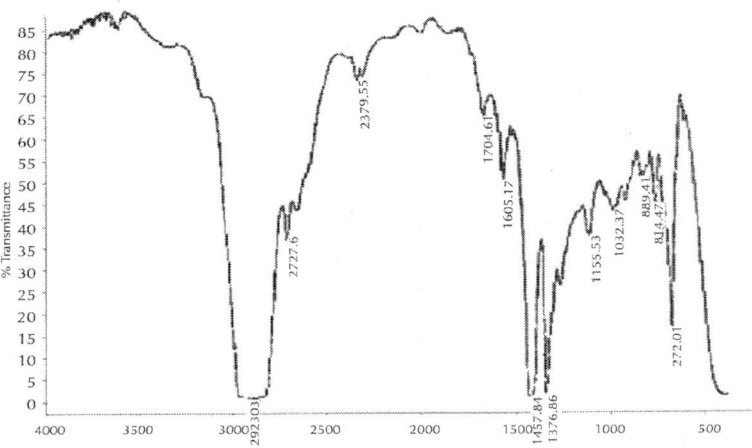

Figure 4: Results of analysis of the used engine oil.

Table 9: Results of analysis of the used engine oil

Frequency cm⁻¹	Type of vibration	Bond	Functional group
722.01	Bending in plain (Rocking)	C-H	Alkans
814.47	Out of plan bending	C-H	Aromatic
889.41	Out of plan bending	C-H	Aromatic
1032.37	Stretch	C-O	Carboxylic acid
1155.53	Stretch	C-O	Carboxylic acid
1376.86	Out of plain bending	C-H	CH_3
1457.84	Bending in plain (Scissoring)	C-H	Alkan
1605.17	Stretch	C=C	Aromatic
1704.61	Stretch	C=C	Carbonyl compounds
2359.55	Stretch	O-H	Carboxylic Acid
2727.6	Stretch	H-C=O:CH	Aldehyd
2923.03	Stretch	C-H	Alkan

IR spectra for used engine oil showed additional peaks at 1032.37, 1155.53 and 2359.55 cm⁻¹ which represent the primary oxydized products at high temperatures. The used oil treated by the acid-clay method still showed some oxidized components as illustrated in Figure 5 and Table 10.

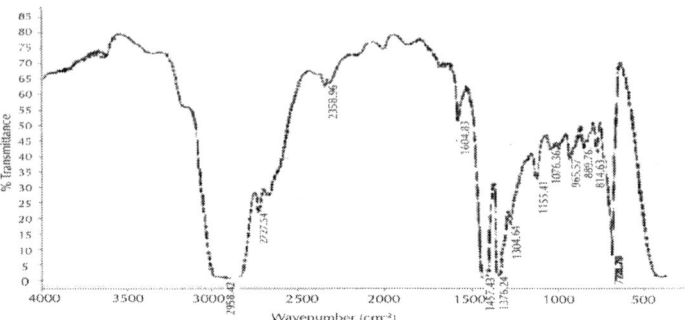

Figure 5: Results of analysis of the recycled oil (after adding the additives) by acid-clay method in local companies in Kurdistan/Iraq.

Table 10: Results of analysis of the recycled oil by acid-clay method in local companies

Frequency cm⁻¹	Type of vibration	Bond	Functional group
721.79	In plan bending (Rocking)	C-H	Alkenes
814.63	Out of plan bending	C-H	Alkenes
889.76	Out of plan bending	C-H	Aromatic
965.57	Out of plan bending	C-H	Aromatic
1076.36	Stretch	C-O	Carboxylic acid
1155.41	Stretch	C-O	Carboxylic acid
1304.64	Rock	C-H	Aromatic
1376.24	Out of plain bending	C-H	CH$_3$
1457.43	In plain bending (Scissoring)	C-H	Alkans
1604.83	Stretch	C=C	Aromatic
2358.96	Stretch	O-H	Carboxylic acid
2727.54	Stretch	H-C=O:C-H	Aldehyde
2958.42	Stretch	C-H	Alkans

There are aldehyde components at 1457.43 cm^{-1} and carboxylic acids at 1076.36 and 1155.41 cm^{-1}. The only difference between the treatment by acid-clay method and vacuum distillation is the absence of aldehyde components as shown in Figure 6 and Table 11. The IR results of the base oil produced by the acetic acid-clay method are represented in Figure 7 and Table 12. This method gives the best results because it shows no aldehyde and no carboxylic acid at 2359.56 cm^{-1}.

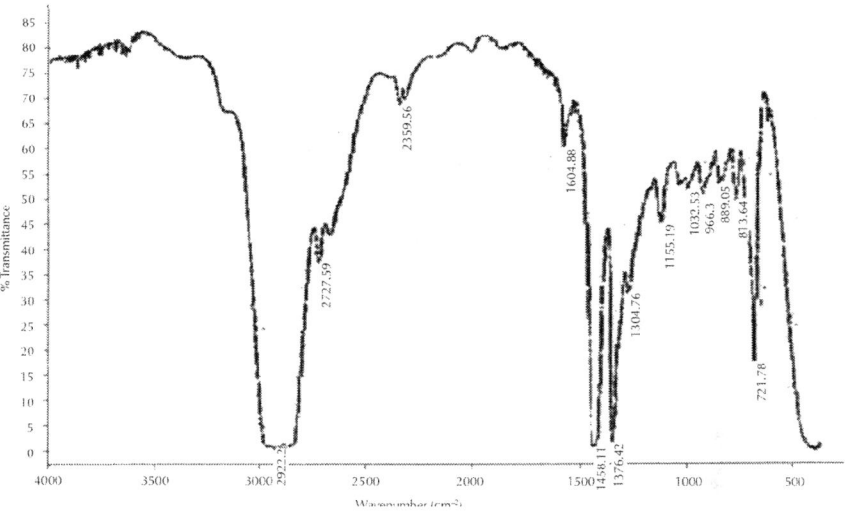

Figure 6: Results of analysis of the recycled oil (base oil only before adding the additives) by vacuum distillation-clay method in local companies in Kurdistan/Iraq.

Table 11: Results of analysis of the recycled oil (base oil only before adding the additives) by vacuum distillation-clay method in local companies in Kurdistan/Iraq

Frequency cm⁻¹	Type of vibration	Bond	Functional group
721.86	In plan bending (Rocking)	C-H	Alkan
813.79	Out of plan bending	C-H	Aromatic
889.56	Out of plan bending	C-H	Aromatic
965.43	Out of plan bending	C-H	Alkan
1032.62	Stretch	C-O	Carboxylic acid
1155.37	Stretch	C-O	Carboxylic acid
1304.69	Rock	C-H	Alkans
1376.94	Out of plain bending	C-H	CH_3
1457.51	In plain bending (Scissoring)	C-H	Alkans
1604.95	Stretch	C=C	Aromatic
2359.56	Stretch	O-H	Carboxylic acid
2727.59	Stretch	C-H	Alkans
2922.2	Stretch	C-H	Alkans

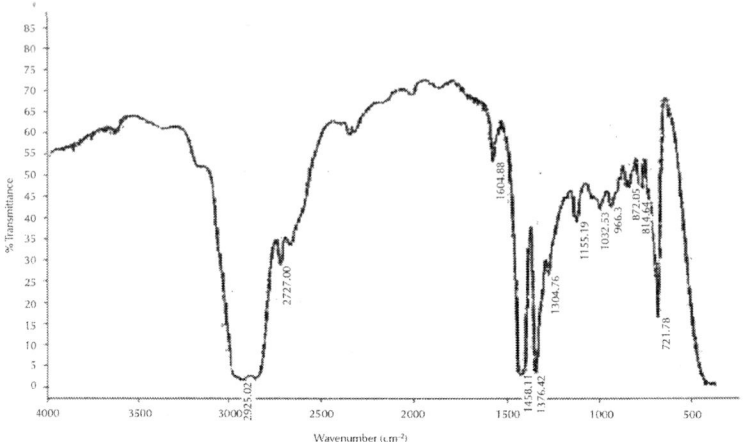

Figure 7: Results of analysis of the recycled oil (base oil only before adding the additives) by acetic acid-vacuum distillation-clay method.

Table 12: Results of analysis of the recycled oil (base oil only before adding the additives) by acetic acid-vacuum distillation-clay method

Frequency cm^{-1}	Type of vibration	Bond	Functional group
721.78	In plan bending (Rocking)	C-H	Alkenes
814.64	Out of plan bending	C-H	Alkenes
872.05	Out of plan bending	C-H	Aromati
966.3	Out of plan bending	C-H	Aromati
1032.53	Stretch	C-O	Carboxylic acid
1155.19	Stretch	C-O	Carboxylic acid
1304.76	Rock	C-H	Aromatic
1376.42	Out of plain bending	C-H	CH$_3$
1458.11	In plain bending (Scissoring)	C-H	Alkans
1604.88	Stretch	C=C	Aromatic
2727.00	Stretch	C-H	Alkans
2925.02	Stretch	C-H	Alkans

There are aldehyde components at 1457.43 cm^{-1} and carboxylic acids at 1076.36 and 1155.41 cm^{-1}. The only difference between the treatment by the acid-clay method and vacuum distillation is the absence of aldehyde components, as shown in Figure 6 and Table 11. The IR results of the base oil produced by the acetic acid-clay method are represented in Figure 7 and Table 12. This method gives the best results because it shows no aldehyde and no carboxylic acid at frequency 2359.56 cm^{-1}.

CONCLUSIONS

This research has shown that used engine oil can be recycled by using glacial acetic acid. This method produces base oil comparable to that produced using conventional methods. Optimum conditions for recycling used engine oil using this method are room temperature and atmospheric pressure. The process for recycling is simple, as it only requires mixing at room temperature, settling, centrifugation and finally mixing with kaolinate. The base oil produced by the glacial acetic acid method is of comparable quality to that produced by the acid-clay method. Also, it has a potential to be reused in cars' engines after adding the required additives. The glacial acetic acid has shown almost no reaction with base oils, in contrast to sulfuric acid, however it reacted vigorously with the used oil. This clearly indicates that acetic acid is not affecting the original structure of the oil. Furthermore, this is most advantageous of using acetic acid in recycling of used oil. This new process of recycling of used engine oil did not emit poisonous gases like sulfur dioxide to the atmosphere. In addition, glacial acetic acid has less of a negative impact on the processing equipment compared with sulfuric acid. Lower amount of additives may be required for the base oil recycled by acetic acid-clay method due its low reactivity with the used oil. Further research is required in order to take this process to the commercial stage. However, while many variables have been studied in this research, there are many others that need investigation such as temperature, pressure, settling time, mixing, centrifugation speed and type of adsorbent.

REFERENCES

1.   Hopmans, J.J. *The Problem of the Processing of Spent Oil in the Member States of EEC*; Report for the European Economic Community (EEC); National Institute for Wastewater Treatment: Dordrecht, The Netherlands, 1974.

2.   Kajdas, C. Major pathways for used oil disposal and recycling, Part 1. *Tribotest J.* 2000, *7*, 61–74. 3.   Boughton, B.; Horvath, A. Environmental assessment of waste oil management methods. *Environ. Sci. Technol.* 2004, *38*, 353–358.

3.   IARC (International Agency for Research on Cancer). Polynuclear Aromatic Hydrocarbons, Part 2, Carbon Blacks, Mineral Oils (Lubricant Base Oils and Derived Products) and Some Nitroarenes. In *IARC Monographs on the Evaluation of Carcinogenic Risks to Humans*; IARC: Lyon, France, 1984.

4.   Francois, A. *Waste Engine Oils Refining and Energy Recovery*, 1st ed.; Elsevier Science and Technology Books: Amsterdam, The Netherlands, 2006; pp. 15–31.

5.   Whisman, M.L.; Reynolds, J.W.; Goetzinger, J.W.; Cotton, F.O.; Brinkman, D.W. Re-refining makes quality oils. *Hydrocarb. Process.* 1978, *57*, 141–145.

6.   Reis, M.A.R.; Jeronimo, M.S. Waste lubricating oil re-refining by extraction-flocculation. *Ind. Eng. Chem. Res.* 1988, *27*, 1222–1228.

7.   Fox, M.F. Sustainability and environmental aspects of lubricants. In *Handbook of Lubrication and Tribology*, George, E.D., Totten, E., Eds.; Taylor and Francis: New York, NY, USA, 2007.

8.   Rincon, J.; Canizares, P.; Garcia, M.T. Regeneration of used lubricating oil by polar solvent extraction. *Ind. Eng. Chem. Res.* 2005, *44*, 43–73.

9.   Rincon, J.; Canizares, P.; Garcia, M.T. Waste oil recycling using mixtures of polar solvents. *Ind. Eng. Chem. Res.* 2005, *44*, 7854–7859.

10.   Shakirullah, M.; Ahmed, I.; Saeed, M.; Khan, M.A., Rehman, H.; Ishaq, M.; Shah, A.A. Environmentally friendly recovery and characterization of oil from used engine lubricants. *J. Chin. Chem. Soc.* 2006, *53*, 335–342.

11.   Martins, J.P. The extraction-flocculation re-refining lubricating oil process using ternary organic solvents. *Ind. Eng. Chem. Res.* 1997, *36*, 3854–3858.

12.   Rincon, J.; Canizares, P.; Garcia, M.T.; Gracia, I. Regeneration of used lubricant oil by propane extraction. *Ind. Eng. Chem. Res.* 2003, *42*, 4867–4873.

13. Quang, D.V.; Carriero, G.; Schieppati, R.; Comte, A.; Andrews, J.W. Propane purification of used lubricating oils. *Hydrocarb. Process.* 1974, *53*, 129–131.

14. 15. Dang, C.S. Rerefining of used oils—A review of commercial processes. *Tribotest* 1997, *3*, 445–457. 16. Beuther, H.; Peno, T.; County, A.; Henke, A.M.; Petreson, R.E. Hydrogenation and Distillation of Lubricating. U.S. Patent 622,312, 15 November 1956.

15. Havemann, R. The KTI used oil re-refining process. In *Proceedings of the 3rd International Conference of Used Oil Recovery & Reuse*, Houston, TX, USA, 16–18 October 1978.

16. Puerto-Ferre, E.; Kajdas, C. Clean technology for recycling waste lubricating oils. In *Proceedings of 9th International Colloquium, Ecological and Economic Aspects of Tribology*, Esslingen, Germany, 14–16 January 1994.

17. ASTM (American Society for Testing and Materials) International. *Standard Test Method for Flash and Fire Points by Cleveland Open Cup Tester*; ASTM Standard D92; ASTM International: West Conshohocken, PA, USA, 2004.

18. ASTM International. *Standard Test Method for Pour Points*; ASTM Standard D97; ASTM International: West Conshohocken, PA, USA, 2004.

19. ASTM International. *Standard Practice for Calculating Viscosity Index from Kinematic Viscosity at 40 ℮ and 100 ℮*; ASTM Standard D2270; ASTM International: West Conshohocken, PA, USA, 2004.

20. ASTM International. *Standard Test Method for Water and Sediment in Crude Oil by the Centrifuge Method (Laboratory Procedure)*; ASTM Standard D4007; ASTM International: West Conshohocken, PA, USA, 2004.

21. ASTM International. *Standard Test Method for Ramsbottom Carbon Residue of Petroleum Products*; ASTM Standard D524; ASTM International: West Conshohocken, PA, USA, 2004.

22. ASTM International. *Standard Test Method for Acid Number of Petroleum Products by Potentiometric Titration*; ASTM Standard D664; ASTM International: West Conshohocken, PA, USA, 2004.

23. ASTM International. *Standard Test Method for Base Number of Petroleum Products by Potentiometric Titration*; ASTM Standard D4739; ASTM International: West Conshohocken, PA, USA, 2004.

24. ASTM International. *Standard Test Method for Refractive Index and Refractive Dispersion of Hydrocarbon Liquids*; ASTM Standard D1218; ASTM International: West Conshohocken, PA, USA, 2004.

25. ASTM International. *Standard Test Method for Density, Relative Density, or API Gravity of Crude Petroleum and Liquid Petroleum Products by Hydrometer Method*; ASTM Standard D1298; ASTM International: West Conshohocken, PA, USA, 2004.

26. Loon, J.C. Analysis of petroleum and petroleum products. In *Analytical Atomic Absorption Spectroscopy: Selected Methods*; Academic Press Inc.: New York, NY, USA, 1980.

27. Skujins, S. *The Analysis of Lubricating Oil Additives*; Varian Techtron Application Notes, 4/70; Varian Techtron Pty. Ltd.: Mulgrave, Australia, 1970.

28. ASTM International. *Standard Test Method for Analysis of Barium, Calcium, Magnesium, and Zinc in Unused Lubricating Oils by Atomic Absorption Spectrometry*; ASTM Standard D4628-02; ASTM International: West Conshohocken, PA, USA, 2004.

29. Lenoir, J.M. Predict flash points accurately. *Hydrocarb. Process.* 1975, *54*, 153–158.

30. Riazi, M.R.; Daubert, T.E. Predicting flash and pour points. *Hydrocarb. Process.* 1987, 66, 81–83. 33. Diaz, R.M.; Bernardo, M.I.; Fernandez, A.M.; Folgueras, M.B. Prediction of the viscosity of lubricating oil blends at any temperature. *Fuel* 1996, *75*, 574–578.

31. Singh, H.; Gulati, I.B. Influence of base oil refining on the performance of viscosity index improvers. *Wear* 1987, *118*, 33–56.

32. Riazi, M.R.; Roomi, Y.A. Use of the refractive index in the estimation of thermophysical properties of hydrocarbons and petroleum mixtures. *Ind. Eng. Chem. Res.* 2001, *40*, 1975–1984.

33. Forsthoffer, W.E. Lube, seal and control oil system best practices. In *Forsthoffer's Best Practice Handbook for Rotating Machinery*, 1st ed.; Elsevier: Oxford, UK, 2011; pp. 347–468.

34. Kishore Nadkarni, R.A. Water and Sediment in Crude Oil. In *Guide to ASTM Test Methods for the Analysis of Petroleum Products and Lubricants*, 2nd ed.; ASTM International: West Conshohocken, PA, USA, 2007.

35. Kishore Nadkarni, R.A. Ramsbottom Carbon Residue. In *Guide to ASTM Test Methods for the Analysis of Petroleum Products and Lubricants*, 2nd ed.; ASTM International: West Conshohocken, PA, USA, 2007.

36. Fox, M.F.; Pawlak, Z.; Picken, D.J. Acid-base determination of lubricating oils. *Tribol. Int.* 1991, *24*, 335–340.

37. Kauffman, R.E. Rapid, portable voltammetric techniques for performing antioxidant, total acid number (tan) and total base number (tan) measurements. *Lubr. Eng.* 1998, *54*, 39–46.

38. Abou El Naga, H.H.; Salem, A.E.M. Effect of worn metals on the oxidation of lubricating oils. *Wear* 1984, *96*, 267–283.

39. Aucelio, R.Q.; de Souza, R.M.; de Campos, R.C.; Miekeley, N.; Da Silva, C.L.P. The determination of trace metals in lubricating oils by atomic spectrometry. *Spectrochim. Acta Part B At. Spectrosc.* 2007, *62*, 952–961.

40. Alder, J.F.; West, T.S. Atomic absorption and fluorescence spectrophotometry with a carbon filament atom reservoir: Part IX—The direct determination of silver and copper in lubricating oils. *Anal. Chim. Acta* 1972, *58*, 331–337.

41. Hopp, H.U.; Erdoel Kohle, E. Atomic absorption spectrophotometric determination of zinc, calcium, barium and magnesium in mineral oil products. *Petrochem. Brennst-Chem.* 1974, *27*, 435–442.

42. Kahn, H.L.; Peterson, G.E.; Manning, D.C. Determination of Iron and chromium in used lubricating oils. *At. Absorpt. Newsl.* 1970, *9*, 79–80.

43. Schallis, J.E.; Kahn, H.L. Determination of tin in lubricating oils with a nitrous oxide-acetylene flame. *At. Absorpt. Newsl.* 1968, *7*, 84.

44.   Bertrand, P.A.; Bauer, R.; Fleischauer, P.D. Determination of lead in lubricating oils by isotopic dilution secondary ion mass spectrometry. *Anal. Chem.* 1980, *52*, 12–79.

45.   Kiss, R.H.; Bartels, T.T. Improved atomic absorption techniques for measuring iron particles in lubricating oil. *At. Absorpt. Newsl.* 1970, *9*, 78.

46.   Peterson, G.E.; Kahn, H.L. The determination of barium, calcium and zinc in additives and lubricating oils using atomic absorption spectrophotometry. *At. Absorpt. Newsl.* 1970, *9*, 71.

47.   Lukasiewicz, R.J.; Buell, B.E. Direct determination of manganese in gasoline by atomic absorption spectrometry in the nitrous oxide-hydrogen flame. *Appl. Spectrosc.* 1977, *31*, 541–547.

48.   Robbins, W.K.; Walker, H.H. Analysis of petroleum for trace metals. Determination of trace quantities of cadmium in petroleum by atomic absorption spectrometry. *Anal. Chem.* 1975, *47*, 1269–1275.

49.   Yusaf, T.; Yusaf, B.F.; Elawad, M.M. Crude palm oil fuel for diesel-engines: Experimental and ANN simulation approaches. *Energy* 2011, *36*, 4871–4878.

50.   Owrang, F.; Mattsson, H.; Olsson, J.; Pedersen, J. Investigation of oxidation of a mineral and a synthetic engine oil. *Thermochim. Acta* 2004, *413*, 241–248.

51.   Maduako, A.U.C.; Ofunner, G.C.; Ojinnaka, C.M. The role of metals in the oxidative degradation of automotive crankcase oils. *Tribol. Int.* 1996, *29*, 153–160.

52.   Von Fuchs, G.H.; Diamond, H. Oxidation characteristics of lubricating oils. *Ind. Eng. Chem.* 1942, *34*, 927–937.

Citations

CHAPTER 1

Walter Holweger (2013). Fundamentals of Lubricants and Lubrication, Tribology - Fundamentals and Advancements, Dr. Jürgen Gegner (Ed.), ISBN: 978-953-51-1135-1, InTech, DOI: 10.5772/55731.

CHAPTER 2

Nehal S. Ahmed and Amal M. Nassar (2013). Lubrication and Lubricants, Tribology - Fundamentals and Advancements, Dr. Jürgen Gegner (Ed.), ISBN: 978-953-51-1135-1, InTech, DOI: 10.5772/56043.

CHAPTER 3

Remigiusz Michalczewski, Marek Kalbarczyk, Michal Michalak, Witold Piekoszewski, Marian Szczerek, Waldemar Tuszynski and Jan Wulczynski (2013). New Scuffing Test Methods for the Determination of the Scuffing Resistance of Coated Gears, Tribology - Fundamentals and Advancements, Dr. Jürgen Gegner (Ed.), ISBN: 978-953-51-1135-1, InTech, DOI: 10.5772/54569.

CHAPTER 4

M. Tauviqirrahman, R. Ismail, J. Jamari, and D.J. Schipper (2013). Artificial Slip Surface: Potential Application in Lubricated MEMS, Tribology - Fundamentals and Advancements, Dr. Jürgen Gegner (Ed.), ISBN: 978-953-51-1135-1, InTech, DOI: 10.5772/55745.

CHAPTER 5

Erik Kuhn (2013). Friction and Wear of a Grease Lubricated Contact — an Energetic Approach, Tribology - Fundamentals and Advancements, Dr. Jürgen Gegner (Ed.), ISBN: 978-953-51-1135-1, In Tech, DOI: 10.5772/55837. Available from: http://www.intechopen.com/books/tribology-fundamentals-and-advancements/friction-and-wear-of-a-grease-lubricated-contact-an-energetic-approach.

CHAPTER 6

M. Shahabuddin, M. Rahman, H.H. Masjuki and M.A. Kalam (2013). Development of Eco-Friendly Biodegradable Biolubricant Based on Jatropha Oil, Tribology in Engineering, Dr. Hasim Pihtili (Ed.), and ISBN: 978-953-51-1126-9, InTech, DOI: 10.5772/51376.

CHAPTER 7

Oghenejoboh K. M. Ohimor E. O, and Olayebi O, Application of re-refined used lubricating oil as base oil for the formulation of oil based drilling mud - A comparative study, DOI: 10.5897/JPTAF2013.0089.

CHAPTER 8

Lisa Starkey Ott, Beverly L. Smith, and Thomas J Bruno, Composition-Explicit Distillation Curves of Waste Lubricant Oils and Resourced Crude Oil: A Diagnostic for Re-Refining and Evaluation, ISSN 1553-345X

CHAPTER 9

Gustavo Pignalosa, Moisés Knochen, and Noel Cabrera, "Determination of Zinc-Based Additives in Lubricating Oils by Flow-Injection Analysis with Flame-AAS Detection Exploiting Injection with a Computer-Controlled Syringe," Journal of Automated Methods and Management in Chemistry, vol. 2005, no. 1, pp. 1-7, 2005. doi:10.1155/JAMMC.2005.1.

CHAPTER 10

Ihsan Hamawand , Talal Yusaf , and Sardasht Rafat , Recycling of Waste Engine Oils Using a New Washing Agent, doi:10.3390/en6021023.

Index